A wildlife students' personal research on Mountain Lions, exploring their behavior, and how they coexist with humans in a ranching, and agriculturally based community. While living out their seasonal range changes on the valley floor, the impacts from obstacle's they face are far greater with human exposure. The research for this project was done in a small eastern part of the Central San Joaquin Valley, of California.

Written and Illustrated
By:
Mikki Terzian
Non-Fiction

In Memory of my big cat, Hookah.

My shadow, my teacher, and friend for eight years.

I miss you every day.

A SPECIAL THANKS

To "SHASTA" THE MOUNTAIN LION
For donating his picture for the cover, and book of
"THE SHADOWS of a MOUNTAIN LION"
&
CRITTER CREEK WILDLIFE STATION
For allowing the publication of his photographs and story.

Please visit Critter Creek
www.crittercreek.org/

SHASTAS' STORY

"A maintenance crew checking high power lines in the Sierras picked up Shasta as an abandoned kitten. The couple running the camp bottle-fed him until they were able to find a home for him at Critter Creek. Mountain lions cannot be rehabilitated in the state of California, so Shasta will stay here. He lets out a peeping vocalization when he sees or hears someone he knows. His favorite toys are pumpkins which he shreds within a few hours. His enclosure is enormous—60' X 40' X 15' at the domed arch. He has tree trunks and stumps, a cave, rock outcroppings—everything he wants to play hide and seek with us. He also has an appetite, eating an average of 5 pounds of meat a day. Mountain lions are the largest cats that still purr and Shasta is the best and loudest at purring."

_____Critter Creek Wildlife Station

Critter Creek Wildlife Station is a non-profit rehabilitation center for wildlife, located
in Squaw Valley, near Kings Canyon National Park. The animals come to us from Fish and Game,
veterinarians, humane societies and caring individuals.

Acknowledgments

This book is for the adventurous at heart. The person who's willing to explore the wildlife that others so often forget exists. To realize the beauty that lives around us in the form of a carnivore, the mountain lion.

I would like to express my thanks and gratitude to Dr. Reginald H. Barrett; Goertz Professor of Wildlife Management, Department of Environmental Science, Policy and Management, University of California at Berkley, for reading my book, lending me advice, and the encouragement to share my experiences with others.

I would like to extend many thanks to my Forestry & Natural Resources Advisors, Kent Kinney, and Wildlife Biologist, Julie Constable, Wildlife Management Instructor, at Reedley College, for reviewing the abstract of my research and sharing their thoughts on improvements with me. I sincerely enjoyed our conversations sharing, and comparing ideas on wildlife and management.

I would like to acknowledge Kim Sorini-Wilson, Wildlife Biologist with the United States Forest Service, on the Sierra National Forest; for the training, guidance, and the opportunity to conduct field research on Forest Sensitive Species'. I learned so much, and aspire to be in your likeness.

Thank you to my family for supporting my notions, theories, and wild jaunts out to some remote place to look for a lion. I have enjoyed the one on one time this research project has afforded us. Many thanks go to the people in our community who supported me, allowed me access to their properties, and helped my project by sharing their stories.

I cannot express how thankful to the girls I am, for showing me the parallels between human motherhood and mountain lion motherhood. This has been the experience of a life time.

About the Authoress

The authoress of: "Shadows of a Mountain Lion," Mikki Terzian, graduated from Reedley College, with an Associate's of Science degree in Forestry and Natural Resources. By completing a work study project for the United States Forest Service, she learned how to apply the knowledge she'd gained through education and personal studies for government required wildlife projects. Today, Mikki is a continuing student completing a Bachelor's Degree in Natural Resources with emphasis on Wildlife.

Studying mountain lions within their own habitat was an eye-opening choice for Mikki. She learned their behavioral patterns through observation, which required patience and intelligent choices to be a cautious human with instincts learned from respect. A six year, personally funded project, rewarded Mikki with "one on one" knowledge that could not be learned through instructional studies alone, she devised her own independent study and subsequently produced answers to questions developed within a protocol tailored to fit the regional ranching and agricultural industries.

Mikki Terzian is a native born Californian, raised in the rural community of Navelencia. The setting of her family's home is in the San Joaquin Valley at the base of the foothills, leading to the Sierra Nevada Mountain Range. Mikki grew up riding horses in the country, and was always amazed at the abundant wildlife.

At the age of 18, Mikki was afforded the opportunity to travel extensively over a period of time to Canada, Europe, and several parts of the eastern and southern United States. Mikki was able to observe nature and wildlife in different temperates, seasonal landscapes, and controlled environments. The larger cities she visited had zoos to observe animals and their behavior within captivity, which provided her with more information about how animals live with relevance to their environmental influences. Mikki chose to do field research on wildlife within its own habitat for the purpose of seeing natural behavior, which brings us to this "labor of love".

It is my privilege to write this dedication to Mikki Terzian. The many hours, the determination to be thorough, and her diligence to her family responsibilities (eight children, throw in a few grandchildren, her parents, and a husband, not to mention, cattle, horses, ducks, cats, dogs, and rabbits). It amazes me she's had the energy to remember all the fascinating facts she's constantly fitting into her compilation of information.

You make me proud, you inspire me, and I hope you have enlightened others as you have me on "this special species' of cat!

Love, as Always.
Mom

Suzanne Hurliman

TABLE OF CONTENTS

TABLE OF CONTENTS Con't

Preface

This book has been a labor of love because it is important to understand how mountain lions behave and why they react to humans, and certain environmental conditions in their own and shared habitats. Throughout all of the phases of their existence; they are in some regions a Key Stone Species', but here in the valley they are an Apex Predator Species'. Additionally, the reader will also be brought to understand their roll and importance in the geographical setting for which they inhabit.

The reader will see how the mountain lion lives within a rural residential community. How they share the same citrus groves as residents in the community that use these avenues for taking daily walks. Overlapping territories of humans and mountain lions do exist and in my experiences, are not as detrimental as most people portray them to be. By reading this book you will have more of an ability to see them as they really are; solitary creatures who've carved out their niche and now in this time have adapted to live within the presence of humans.

For the Ecologist, Wildlife Biologist, and Wildlife Student, I believe this book is valuable because it shows how the co-existing mountain lions live and survive in a predominately high intensity agricultural area. The seasonal activities and range changes of these mountain lions listed on the family tree have been documented by sight, track, and sign, and by frequency of travel with fixed points. By reading further, their stories will come together. I hope this research lights a fire in wildlife students. Though the work is time consuming, it is rewarding. We must be their voice.

If the reader is not a scientist or science oriented student, he or she will find each lion's story to be heartfelt and informative. By experiencing the effort that each lioness extends in raising her cubs, you will see that a lioness is family oriented. She will do what is necessary to protect and raise her cubs from their birth until the day they become subadults that seek their own home ranges. My field notes show that human motherhood often parallels mountain lion motherhood.

I never intended to write a book. An automobile accident left me with a spinal injury that required time to heal. Writing this book was the perfect opportunity to share my research and keep my mind busy.   Between my family's activities, school, work, cattle, illness and injury, this project has been a huge personal effort. My project is not funded by any program(s). All of my equipment, I purchased myself.   I've done radio telemetry in the past but; this was not an option for my own research.

Today, mountain lions are collard and radio telemetry is used for tracking studies. The data is GPS recorded with computer software for future analysis. It makes it easier to follow mountain lions routes when they are in topographic areas where tracking is difficult.   However, in some ways the old way of doing things is better from the learning stand point. These journal entries are taken from my field notes and only the best entries are listed. The four study zones encompass several miles, and are more or less in some areas, connected by waterways.

For the purpose of anonymity, I have chosen to change the names of people in order to protect the innocent. Any resemblance, living or dead, is entirely coincidental, except for the names of my five children, husband, parents, and referenced works and citations. I have chosen to share the names of three locations for the sake of studying mountain lions. The mention of these locations excludes any private or government environmental studies or impact reports. They are: The Friant Kern Canal, Kings River, Wahtoke Lake, Campbell Mountain, Jesse-Morrow Mountain, and the San Joaquin Valley. I will also, not show maps with fixed points or routes that each lion has in order to protect them.

Under a clear blue sky, come and sit with me on a mountain top overlooking the landscape. Feel the gentle breezes on your face as the grasses sway around you. Hear only the sounds of nature; feel like a lion with me.

Chapter I
Discovery

In 2002, my first experience with a mountain lion sparked my interest and my passion for studying carnivores. Just seeing a lion was not enough, I wanted to know where it was going and where it had been. Could this lion be a mother of cubs or was it a male surveying his home range? I had to find out. Taking walks turned into a past time for tracking wildlife. It is interesting to see where lion's tracks will take me. I have found evidence, by track, of mountain lions stalking their prey. On other occasions, I've found lionesses' tracks along with their cubs as they've traveled to different locations, and seen evidence of their kills to feed their cubs.

Then, the shocking occasion where I have accidentally lost the tracks only to look up…to stare into the eyes of a carnivore. My heart stopped for a brief moment, only to feel it beat again quickly. I was able to slow it down, knowing that at this moment every movement counts.

Not every encounter is the same. Where one may not be confrontational, there may be another that is. A chance sighting can happen so quickly. When it has occurred in the past, I've read the cat's body language and the pupils of their eyes. I make sure to keep the eye contact, not to lose a second of it. By observing the lion's movement or stance, and watching for tail swishing, I have been able to coexist with the resident mountain lions. Tail swishing can be construed as aggressive or nervous behavior. When encountering a lion during hunting hours, tail swishing can be seen as an aggressive behavior. The lion is calculating its preys' behavior for an attack. When accidentally running across a lion while walking during the late morning hours, the swishing can be observed as a nervous sign. The reason for this explanation is because the lioness I had observed at close range was obviously not prepared for a meeting with a human. This lioness had made her way up the backside of a rock pile in search of a place to nap with her cub.

I asked myself; can a mountain lion in the study smell fear from a human? When I met up with Suka, I found that a mountain lion will look for physical movement which will indicate how it should make its next move. I know that when I'm surprised by a person or an animal that my heart beats faster, adrenaline pumps through my body. This might not be a good thing when standing in front of a carnivore. My glands will produce a scent that the animal probably isn't familiar with being they are wild. During the closest chance meeting I've had with a lioness, I found that if I'd been upwind of her, my scent coupled with erratic physical movement like that of a nervous prey species' would've lead to danger. Luckily I was downwind from Suka, and she couldn't catch my scent of surprise. At the location of our meeting, the wind current flows from South East to the North West before 6:00 p.m.

Perhaps the interest I showed in Suka made her wonder why I was still standing there. It seems the pupils of our eyes never left each other's for more than a second as we evaluated each other. Her eyes were as wide as mine yet, not expressing any fear from the meeting. I had seen lions from a distance, followed their tracks, and found their scat. There were times I'd wished I could have seen them closer but, not face to face. There was no amount of close up pictures that could have prepared me for seeing the real thing so close. She was beautiful, so self-assured, and inquisitive. If she'd had a voice she probably would have said to me, in a raspy tone, "Well, now what are you going to do?" With the way she looked at me, her piercing yellow eyes looked right through my soul. I couldn't help looking back at her with wonder to see such a beautiful cat. She stole a part of my heart that day. Since then, I have looked for her everywhere and on many occasions found that I've followed in her footsteps.

Though our footprints are different, the way we walk through life is similar. I say this because she and I both have a small child to raise. We feed, spend time, protect, and nurture them. We mother's get few and far between moments of quiet time. But then, when we least expect it, a brief respite comes our way. We can look out over the landscape and have peace. Who knew that this day, both of us would meet? What a surprise it was when our eyes quickly focused on each other's. As two different Species', we discovered one another and I came to realize the parallels of our lives.

Suka taught me that upon leaving a lion's presence, to keep backing until I have enough room to turn sideways but never let my guard down and never turn my back. I knew when she looked at me that she would find my departure interesting. Since her

14

nature is to track her prey I knew she may decide to track me for a short distance. I also knew not to be alarmed at this, all cats do it. She just had to make sure I was leaving and really, she wanted to see if I would run. She would love to chase me and wouldn't be able to resist the opportunity. I love the interaction with these wild felids, but I also know that a chance meeting should be controlled, brief and distant.

From this time on, I followed a mountain lion and there were times when she too, followed me. In some ways it became a game to see which one of us could detect the other and yet stay obscured, as not to be seen by the other. I would hear the occasional faint rustle of leaves, or see fresh tracks from her and her cub. I knew she was there and she knew I was, too. I visited the same area at different times of day, every few days, to see if I could find her. Most of the time, I was not disappointed. Even if she had decided to leave before I could look for her, I always knew the direction she left to because she did not hide her tracks. Although, I have found that a lioness will hide her tracks when a male is in the vicinity for the safety of her cubs. They do this by walking through the debris of fallen leaves, if they haven't left the area all together.

A similarity between humans and mountain lions is that we both protect and feed our young. Hookah is the largest lioness in the area. Her hunting skills are unsurpassed. The dogs may bark a little while they suspect her presence. She waits in the shadows.

Late at night, when she's in the area, I can hear her when I listen from our bedroom window. The dogs do not creen with her presence as they do with the younger male. Perhaps it is because he makes his presence blatantly known. Hookah remains in the shadows of the trees just down the hillside. The dogs become silent, waiting for an indication that she's here. The crickets stop chirping. The dogs hide. The silent huntress knows it's now her turn to make a move.

Hookah has a few different strategies for her hunting, and the sounds she makes could fool anyone into thinking she is a kitten in distress, possibly stuck in a tree. In darkness, Hookah hunts the back acreage just on the rim of the orange trees. The species' which is most prevalent is the one she hunts first. She may sound like a bird, or a screaming rabbit. There is a method to her screaming rabbit imitation. Within each zone of habitat lies a small section of land where water runoff has created the perfect hiding place with tall reeds for coyotes to hide. I have followed their tracks and seen where they enter their grassy tunnels. She will find them if they're there. They cannot hide from her.

Out of sheer amazement of the mountain lions' beauty and nature, began my goal of species' specific field research. Though each predator does a job to keep the ecosystem in balance through aid of another, I see that our mountain lion keeps the coyote population at a manageable number. In turn, with the presence of the mountain lion and the threat they pose to the coyote, the coyotes keep moving even throughout the day. When the coyotes are not focused in one or two major areas in number, they are able to keep the rodent populations down because they cover larger spans of land. I've seen them traveling by pair or a small group of three to five. They seek the cover of orange groves during the day as they hunt rodents. They keep moving to avoid the morning hunts of the mountain lion. They can slink through the orange trees all they want to. The thick foliage in reality is only a false sense of security for them. The lioness watches from a hill top to see the directions in which they travel. The small open space between each tree provides a periodic glimpse of their activity and direction of their travels, to the observant mountain lion. She waits until they rest.

I could not imagine life without the adventure and education that this animal provides me. I hope that my shared experiences help people gain an understanding of the mountain lion's behavior within their own habitat. This understanding can help to indicate why or why not, an attack may occur. Human activity in some cases could be construed as provocative and escalate the mood during an encounter thus leading to further issues. It has been my experience that once an initial action to a single lion has been made, their reaction along with the humans becomes learned. I have seen that it only takes one time and they never forget.

When my husband and I go camping we prefer to choose secluded places. When we find a great spot away from people the next thing that goes through my mind is that, the least amount of human traffic in that area makes for better lion habitat. Signs warn people about the dangers of mountain lions, but no one ever thinks they'll see one. The mountain lions here within the study know what people are and try to avoid them. However, mountain lions at higher elevations may not know people. In some cases, subadults are so curious and unafraid of people that they put themselves at risk of being killed by being seen. This is where humans come in. I feel that it is not only my responsibility to keep the mountain lions safe by keeping their locations concealed but, to keep myself safe so that there are no bad encounters.

Mountain lions are instinctual, but it would be a mistake to think that they do not reason almost as well as a human can. Keeping in mind that people have more power to reason than most species', it is better in the long run to scare them off so they will remember that alarm the next time they are in front of a human. It may save their lives.

When recording information in a journal it is customary to give the date, wind speed, direction, and note weather conditions in each entry with a summary of events. Not all of the entries in this book are noted this way in part, for the sake of telling the stories. I have more pictures of these lionesses and their cubs than I thought I would. The only problem is that they're either not clear or too dark. A mountain lion has a way of becoming almost invisible with the way they blend into their surroundings. I am always amused at the pictures. When I pan across the mountain chains, I can't always see them because they blend into the scenery so well. When I get home and download the pictures into my computer, I often see a lioness and her cubs peering back at me. Through the shadows created by tree limbs overhanging large boulders, a surprise waits for me. The shining reflections of the shapes of their eyes, the faint outlines of their heads and bodies, the apparent spots on a cubs back, and a long tail draped across a rock, is confirmation to me that they've survived another day.

Chapter II
The Beginning

The majority of my life has been spent in a small rural community at the base of the foothills. Country living is my preference. The nearest town is ten miles away. The foothills all but surround a moderately sized natural lake that connects to other riparian areas. The wildlife is abundant here. When the lake is full there are countless ducks that cover the waters' surface. While sitting on the bank of the lake, the sounds of bull frogs, ducks, and insects fill the air. If I sit really still, I feel as though I've become part of the landscape and nature just happens around me. The finishing touches are the grasses and reeds that sway in the gentle warm breezes. Even when there is a drought the wildlife finds a way to survive. The tops of the foothills are just as serene. Looking down at the valley from a higher elevation gives perspective to why a mountain lion would choose this location over all.

Many years ago I moved to the foothills. I knew the place was great habitat for mountain lions but, didn't really give them much thought. Like other people, I paid no attention to the fact that they could or would come to my house. One evening during the summer, I painted the concrete on my back porch. Later that night I had been sitting at the computer with my back to the French doors in the dining room. The porch light was on. I figured that was enough light to keep the raccoons, and pigs away. The next morning I got up and opened the French doors to let the morning air in. When I looked down to admire the fresh coat of paint on the concrete, I became rather irritated when I saw several tracks across the porch, all through the paint.

This is the panoramic view overlooking the Ag land from the top of a foothill in zone 3.

As I paid closer attention I could see that they were not the tracks of a large dog, which would have been the best possibility in my mind, yet realized they were the tracks of a mountain lion. As I looked around my backyard which consisted of a courtyard behind an open ended retaining wall, I thought, how did this happen? I left a light on. The dog should've been barking if there was something outside. The dog didn't bark, probably out of fear of becoming a meal, and the light only helped the big kitty see where he was going! The next thing that went through my mind was, "Where is it now?" Looking around the house I could see nothing out of the ordinary. The pigs were visiting down in the creek bottom rooting away, the German Shepard was asleep, the rabbits were waiting to be fed, so were the horses. Everything was in order. Everything but the cat tracks through my fresh paint. Then I thought to myself, "When could this have happened?" Apparently when the paint was still wet! Somewhere out there in the morning sun was a mountain lion basking on a rock with beautiful bluish, gray feet.

Now is the time that I had become worried enough to call the Department of Fish and Game. The reason I'd called was to report the incident like I had always been told to do. The California Fish and Game Code Handbook under section 4800, says, "The mountain lion (genus Felis) is a specially protected mammal under the laws of this state (Lexis Nexis Gould

Publications, 313)." I believe that in 2011 the genus was revisited and changed to Puma, but at that time Felis was current. Nevertheless, a biologist came out to visit soon after I called. I was glad to see him in hopes he would issue me a permit to take the lion, should I need it. Looking back, I think the reason behind my thought of needing a permit was only for the fear of something out of the ordinary and the lack of understanding. There were no signs of the lion stocking my children, I hadn't seen it. Seldom does anyone see this coming, but an animal that is stalking your home is actually pretty blatant about it. If this was happening, there would have been more tracks around the house, and a possible sighting of the lion watching the house. This hadn't happened. The animals and stock were fine and really, I never thought we could be that interesting.

The permit idea was denied and for good reason. Rabbits are a prey animal, and I had two. The area surrounding the house was all natural habitat; the topography was uneven with slopes and draws, several large boulders, rocks and tall grasses. There was a running creek down below the house with a seasonal pond at the top of the mountain. If a lion were walking across the property it wouldn't be doing any harm. My last dawning thought was about my noisy four year old son. This child of mine could shriek loud enough to hear a mile away, and we all know that couldn't possibly raise the attention of any wildlife! In my mind at the time, I thought all wildlife would have vacated due to the noise. However, looking back I remember that mountain lions are drawn in by the sound of squealing prey. This is where noise from the children comes in. They had high pitched voices that attract mountain lions in search of a meal. Listening to the sound that baby pigs make when they lose their mother or try to catch up to her, reminds me of the way the girls would squeal out of excitement. They also used to tease the pigs this way, at least until they were all chased up into the trees.

To summarize every reason that the cat found us interesting is because of the smell of food cooking, and small game in close proximity of the house, the riparian area, the squealing of my kids and their friends, the momma pigs with babies crossing the property, and the possibility of the poor thing needing to get a drink of water. After the biologist left, I thought of these things a bit more seriously, and was actually very surprised at myself for not recognizing this in the first place. The only thing to cross my mind at the time was the safety of my children, and the taking of a lion.

That is the protective side of being a parent. A lioness would have thought the same thing if the event were reversed. However, humans have more options. I thought to ask myself if, in the surroundings that I had chosen for myself and my children, within a lion's habitat, had I forgotten where I was living? Or, did I take for granted that because other people lived in the community, that wildlife would step aside? Then at that moment, I thought, "How can they step aside for people in the community when it was people that moved into their house?"

Later, I did see the mountain lion. It could have been male or female; I wasn't sure at the time. Now, I do think it was a young male lion with large feet but, I could not see if there were any faint yellow spots on his coat to give him an approximate age. The only thing I saw was the happy lion leaping and bounding over the tall grasses after something that had sparked his interest. That was truly a peaceful moment. Later the following year my children and I moved down to the valley near the base of the foothills where I grew up. This has been the setting for my study of Mountain Lions. When living at the base of the foothills in a rural farming community, I think back to when we lived just a few miles away in the foothills. Since then, more people have moved up there. The increasing population over a short amount of time has pushed the mountain lion population down closer to the valley floor. They are being displaced from their natural habitat by humans. Where else are they to go?

Chapter III
The Morning Mist

Before my new husband built our home, we would periodically stay here at the ranch in a bunk house. We used to sit on a little porch in the early mornings, drink our coffee, and watch the sun come up. In early fall of 2006, mist floated over the creek bottom. As we sat with our coffee, I saw a long figure emerging through the mist. The animal slowly advanced up the hillside and out of sight. It was a lion; the long tail gave it away. The look on Tracy's face said it all. He was amazed. Tracy was in denial about it being a lion even though he'd watched it the same as I had. I still can't say which lion it was but I have a feeling it was the gray lion. He blended into the mist so well. We didn't see any more of the lion and before we knew it, winter was here. There were times when I thought I'd seen something moving along in the fog. I wasn't quite sure what it was. I figured it could have been a coyote. The seasonal range changes for Barbara and Sneaker happens in the late fall, and early winter.

The winter of 2008 came. As time went on we really hadn't thought much about the lion. Then, one morning my daughter Mel had been warming up her truck to go to school. She came in the house and told me she had gotten a weird feeling when she'd seen something moving through the morning mist. She said that it didn't move like a coyote. I told her it might not have been. The fog in the Central Valley varies in ranges of visibility during the winter and early spring if there's fog. Sometimes it only allows for a visual within a short distance. The larger wildlife uses the fog to their advantage. An animal can be one to two hundred yards away from you or sometimes less, and you can't see it. Barbara Alice and the other resident lions are the reason that the coyotes keep to themselves throughout much of the year. Two lions specifically, hunt here in the fog. Coyotes normally don't call to one another in this weather condition. They save that rotten behavior for clearer evenings or colder nights during the winter. In the summer evenings, depending upon when, and which lion decides to venture through, the coyotes seem to have a continuous party.

In the spring of 2008, the grass was thick and there was still a chill in the air with dew on the ground. I had just had my last child in February. The month was now March. With the help of my girls to watch their new sister, I was able to get out of the house for a little while. I drove to the back of the lake where I had recently seen deer. I went out to set up a camera. There were no cattle in the pasture; no Red Tailed Hawks in the tree or any signs of wildlife. I knew something had been there though. The large, trampled down patches of grass gave away the earlier presence of deer. I grabbed my infrared camera and scanned the area for a good location to set it up.

The morning mist was still in the air as I scouted for a tree that would both hide the camera, and still allow the lens visibility. The farther away from my truck I got, the more odd the silence felt. I thought I might be a little silly to be feeling like this. My senses seemed to be heightened. The skin on my arms was more sensitive as I walked alone. I rolled my sleeves down thinking that might straighten things out. A few minutes went by, and I had found a tree at the back of the pasture. I took my time setting the camera, enjoying the peace of the morning. After the camera was set I sat still by the tree. I looked around in every direction. There were no small birds chirping, the morning was still while the fog rolled through. There was dullness in the air. The echoless sound of dew dripping from the leaves was the only sound I'd heard next to my own breath. It was time to go.

I began walking back to the truck. I knew I had company. It felt like there was a pair of yellow eyes trained on my back. My experience tells me that my body knows when something is out of the ordinary even when my mind hasn't figured it out yet. With every step through the thigh high, wet grass, I felt a presence right along with mine. I thought to myself; if a lion were here, would I not see the grasses moving? Would I not see an indication of any kind? Then it dawned on me. There is cover that lines the empty creek bottom. The fallen tree branches, thistle and other tall dead weeds from the season before, made it possible for a large animal with furry feet to stalk me without my knowledge. The sandy bottom of the creek afforded the perfect opportunity for a lion's secrecy.

Upon this light bulb now shinning bright in my head, I felt the cool air rush over my skin. Very soon I would have to cross this creek bottom in order to get back to the roadside. I still had some distance to walk before I reached the bank. Keeping a slow, even paced walk, I paid special attention not only to the twigs I was

23

snapping in the wet grass but to the sounds near me that were faint. I knew if a mountain lion were in the empty creek, I'd never be able to see it the way it could see me.

As I neared the place where I had to cross, the blood was pulsing through the veins in my neck. I stopped at the bank side to have a short look around me. I felt the silence become louder, and the feeling of my space being intruded upon from behind my left shoulder become stronger. There was no movement, just the dull sound of dew floating by. Perhaps my company had stopped when I stopped. It became very clear that my company was stalking me. With this in mind, I cleared my throat with a grumbling low tone and stepped down into the sandy bottom of the creek bed. For some reason this weak attempt to sound bigger than I was made me feel a little better. Prepared for what seemed inevitable, I braced myself a bit for the short distance around a fallen tree, then back up the bank side onto the bank. The farther from the creek I got, the better I felt. But, I wasn't in the clear yet. I was not stepping on twigs of old iron weed anymore yet; I could hear a faint rustle in the creek bottom as I gained distance. After reaching the fence line and the safety of my truck, I sat and watched. Though I saw nothing emerge from the creek bottom, I knew something was still there. Perhaps I'd gotten too close to a deer cache and didn't know it. Had the curious visitor found me as interesting as the deer that had been there earlier? If so, I probably wouldn't be here to tell the story. Luckily, the only interest I'd provoked was that of curiosity.

The next morning, I went back to look for tracks along the creek bed. But, this time I didn't go alone. There were more indentations in the grass indicative of a number of deer in the herd. I was prepared to find, "the something," that had stalked me. To my dismay I found nothing. Then after searching a bit more down in the creek bottom, I found the tracks of a mountain lion. The tracks followed the creek bottom in two directions, and tracks were all over the bank. Sand can spread a track out to look much bigger than it really is, but it didn't matter. These tracks were big anyway.

Two weeks later, I went back to the same site to collect my camera. The overall mood of the wildlife was busy. The sun was out, the birds were singing, all was normal. Nevertheless, I watched the creek bottom carefully while I thought about my prior visit. I'd come too close to seeing a mountain lion while it hunted.

While researching methods on tracking, I came across an article on tracknature.com. The article was written by James C. Halfpenny Ph.D., the title of the article is, "How big is that Track?" His information was derived from Fjelline, D.P. and T.M. Mansfield, "Method to standardize the procedure for measuring mountain lion tracks," which he makes reference to in his article. Dr. Halfpenny's work can be found under, Snow Tacking, in chapter five of, "General Technical Report PSW GTR-157, American Marten, Fisher, Lynx, and Wolverine: Survey Methods for their Detection." The information I've used for tracking is the method of minimum outline and variable outlines of a mountain lions' track. This is extremely valuable information to me. An illustration in Dr. Halfpenny's article helps to visually explain the difference between the minimum outline and the outside, variable outline. Dr. Halfpenny also gives a great example of how to use this method. By following the written instructions I was able to differentiate between all of the lions in my research zones. After becoming familiar with this method, I didn't have to use it all of the time unless; a new lioness entered my research area. It came in handy when I needed to make sure which mountain lion was which, and why they were there in the study. What an asset this method is for identification.

The pictures from the camera revealed plenty of deer, but no lion. However, there were pictures of darkness. I have tried to keep a record of the pictures that turn out this way for the simple reason that it can help me establish a time line for a lion's average nocturnal hunting hours. This of course needs to be backed up with track or sign to double check my theory. Sometimes it works. If I were to speculate on which lion it was that graced me with its presence, I would have to say it was Sneaker. He enjoys deer meat and there were plenty of deer here to eat earlier that morning. The tracks he left were identifiable on the other side of the road in the soft top soil of a citrus grove.

Chapter IV
A Tiny Family

School had started again in August of 2008. Everyone was in the house getting ready to go. As usual, Tracy and I were having coffee on the front porch. He was about ready to leave for work. We were talking about our plans for the day, when something caught my eye across the road. I said, "Oh wow! Look at that, it's a lion!" Tracy said, "No, there's no lion." I said, "Yes! Look!" After he focused on where I was pointing he saw her slinking down the hill, passing under the almond tree to stop once and look toward our direction. Then she continued down into the creek bottom. I was so happy; she was the most beautiful cat I had ever seen. Her coat was so creamy white that it shown in the sun. She moved slowly with such grace. She was in no hurry, and knew we were on the porch watching her. Tracy looked at me in disbelief for what he was seeing. Like most people that live on the valley floor, who would think that with all of the farming that happens here, we would see a mountain lion. The feeling I'd gotten from seeing her was inexplicable because I knew moments like this were so rare. While watching her as she chose her footing so carefully, it felt like an awakening. Out of another ordinary day came a surprise that I continue to look for today. Hookah was the topic for quite some time after that, and we made sure to look outside before opening the doors.

Hookah was not a large lioness at the time. Her standing height was about three feet tall. The elevation of this sighting was at 397 feet. After seeing her so close, I would go out on the porch in the mornings with my coffee, hoping to catch a glimpse of her. When not seeing her and knowing she was there, I started taking walks in mid-morning with a wolf that someone had dropped off at our house. Stormy was a hybrid, he had blue eyes, and I think he was crossed with malamute. We found scrapes in the creek bottom from where a lion had previously marked its territory. I also found scratches on an old tree at the back of the property that defined boundary marks. No doubt that these belonged to either Sneaker or Hookah. At this point in time we had cattle. The fences were as usual, in need of some repair. When Tracy went out some mornings to mend fences after coffee, Stormy and I would take a walk out to see him. One morning around 8:00 a.m., he said the field worker had seen a big cat. She was sitting on the pond bank just watching him. I asked if the man had said which direction she'd gone. He told me she went back down into the pond. The wolf was beneficial to me because what I didn't see or hear, he did. It was easier to watch him for indications of her presence while I tracked her. The wolf was my early warning system. Thankfully, his hearing and eyesight were still good.

Our walks became more frequent. I would find scat and there were poofs of feathers everywhere. Hookah was a birder. The weather had changed, and three days of rain came. I really wanted to get back outside and continue tracking her.

Learning about what she does during her day was important because I was lucky enough to have her within walking distance to learn from. Finally, the rain stopped. The ground began to dry out. The soil in zone 1 is a silty clay loam, thus when tracks are created by animals in this soil type, they are well defined and preparing casts is easy. They turn out nice. I do have to make sure that the ground has had ample time to lose at least half of its soil moisture content. It helps to mix the Plaster of Paris somewhat dry in moist soil.

Another excursion a few days later led me past the dry pond reservoir. I started to feel like I had grown eyes in the back of my head. I had to keep an eye out for anything moving, while looking for tracks. I worried about the big kitty popping her head up from the top of the bank. Little did I know how very busy she too, had been. After turning the corner away from the pond, I was walking by an orange tree when I found her tracks in the wet soil. They were not only her tracks. Hookah had given birth to a litter of two cubs. The cubs were now at a stage where she could take them out with her. On the wet ground near an orange tree, I'd found the tracks of the tiny family. Hookah had walked around in a circle before she sat down. Her two cubs tracks had circled around hers, leaving their tiny foot prints all around where she'd been sitting. I wondered how long they'd sat there, and what they'd seen. The tracks led off down the access road of decomposed gravel into the tall grasses, and down into the creek bed. The cub's tracks never strayed from their mother's side. One set of tracks were slightly larger than the others. From this I gather that one is male, and the other a female. It was cute to see how the cub's tracks tangled around their mother when she was sitting down. These tracks were so small that I had to look closely to see any differences.   I decided to call them Ben and Snuggles for the purpose of telling them apart.

Looking back, I can remember that in the month of May, we had heard screams outside. I thought they were from an owl, hunting in the darkness. They were high pitched, and only a few that I can remember. I now know that these screams were indications of a female in the creek bottom. Considering the timing of the screams to when we saw her across the road, it's safe to say that she had mated and come back to the area to give birth to her cubs. Sneaker had been here. We never saw him but, I did find the scrapes of debris with scat on the tops of them. This was not only a boundary mark but, a calling card to let Hookah know he was there.

Chapter V
Survival

In September and October, we had cows popping calves out all over the place. Hookah's cubs were old enough to go hunting with her now. It was funny to drive by the upper pasture in the mornings because all we could see were little white or gray poofs of feathers or bunny hair. I can just imagine how much fun she had teaching her cubs how to catch their prey. Up by the grave stone at the corner of the pasture, rabbits and squirrels have made their homes in burrows under the base of the stone. They come out to feed at night and in the early morning. The birds start flying at the break of dawn. She had her cubs out hunting with her at four in the morning just before the sun came up. Most of the human residents don't drive down this road until 6:00 a.m.

In the evenings when we fed cattle, we counted calves to make sure they were all there. They were, and life was good. I did have to wonder if that would last because I've heard that as a lion gets older they develop a taste for beef. With age, they may change their hunting habits because they develop more skill and become bolder. Prey availability during drought seasons also plays a role. If the available food source is low, lions will hunt the next favorite food item on the list. Nonetheless, while she was here she behaved quite well. I was glad for the good experience and the opportunity to learn from her and her cubs.

Hookah has some sense to her. We have a top pasture by the road. It's fenced and the cows couldn't get in unless we opened it up for them. This is where she kept her cubs when they hunted in the mornings. She knew the cows would've gone ballistic, and smashed them all into the ground had she taken them down into the lower pasture. I saw the way the cows immediately clustered together in excitement and tried to stomp on a few of my house cats when they'd follow me down there. That was short lived. The cats figured out they should stay in the yard. Our cattle were funny. It was like they had herd masters in the group. Birdy the Brangus was the head of their organization. She'd give out a bellow or grunt and the others would come running. Janice was the translator. I'd yell at her to come on, and she'd turn her head, let out a terrible noise, every cows head would pop up, and here they'd come. The calves all made it through Hookah and her cub's visits just fine. Our worry was coyotes, wild dogs, and dust pneumonia.

If a lion were to kill a calf or ten calves, the rancher has or should have insurance that will cover herd loss. Between a mountain lion and dust pneumonia, the chance of a rancher loosing calves to dust pneumonia is far greater every fall, than if a lion were to take a calf. This is especially true if the rancher has a large ranching operation. The chances of finding all of the sick calves are slim yet, possible if they go out every day. There will be loss no matter how a person looks at it. The name of the game with cattle is to try to prevent any loss for future monetary gain. However, most cattlemen experience loss with cattle no matter what the loss is attributed to. I can see losing one or two calves but, after that the rancher should consider getting some goats. I don't believe that feeding a wild animal is the answer but, I know for a fact that it works as a deterrent. The worst case scenario is that a habitual offender moves into an area because a lion has died, and his niche is opened up to others. The new lion will either eats the goats or ignore them, and goes straight for the calves. Mountain lions are accustomed to eating deer. Being pushed down in elevation into more cattle grazing land might be part of the issue with ranches and residents in the foothills overcrowding the mountain lion. Not all ranchers seek to take a lion. There are some though few, that wait to see if there is loss attributed to predation. On the side of the rancher, it is a huge concern when loosing cattle repeatedly to one mountain lion. They do not always take the sick, old or week. At this point one needs to consider whether an adequate prey base is available and what may have caused a possible decline in prey. When one species' under the mountain lion is out of balance there is an affect.

On a dreary day in October, I had to run to the store for groceries. At the top of the hill, off to the left side of the road is an orange grove. It was around 1:00 p.m. when I was driving past. I looked down one of the dirt access roads and to my surprise, saw a small yellow cat with a long tail. It was to say the least, in a bind. Two large coyotes were playing monkey-in-the-middle with the young cub. As I watched from the road, the cub had nowhere to go. In an effort to free its self from the immediate danger, it ran with its tail flying down through one of the citrus rows. I was really worried about the outcome. To my dismay I never saw both

30

of the cubs again. From then on, the only tracks I would see were of Mom and one cub. Out of the two sets of tracks, the larger remained. They were not Snuggles tracks, but the tracks of Ben. Periodically, I would still see Hookah sitting on the pond bank. By her far off gaze out over the still pasture land, I wondered if she'd thought of her missing cub. A lioness and a human mother are comparable with the love and devotion that they have for their cubs, and children. Time does not stop even for the loss of a lion cub. One must eat.

Hookah's love for birding nearly got her in trouble with the S.P.C.A. Mr. Savage had, and I stress, "Had," guinea hens. They were in a pen in his yard. The man was exceedingly upset one day about the neighbors' dogs getting into the guinea pen and killing a few of them. However, he had never seen the dogs in the pen, and no dead bird's remained. There were no signs of the guinea hens except for a few loose feathers on the ground.

Disagreements between Mr. Savage and the dog owner's grew to the outcome of Mr. Savage calling the S.P.C.A. out to his house. While Mr. Savage was complaining about how the dogs spent time in his driveway, the agent from the S.P.C.A. told the man to look at the tracks around the pen. As he did this, the agent told him that those were not the tracks of a dog but, the tracks of a cat, and more than one. The man then started on a rampage about a bird steeling lion.

After a few weeks had passed, I spoke with his foreman who then drove down to our house to look at the roof over our duck pen. I told him it helped to have a roof on the pen not only because of large cats, but owls as well. I asked him if their guinea pen had a roof. Perhaps the man misunderstood me but, he said, "No." While we were talking, Mr. Savage came down. He wasn't in a particularly good mood, and especially after I told him he may want to consider adding a roof to his pen. He said it already had a roof, and got back in his car to leave. As he was leaving, he stopped and swore up and down he was going to catch that cat and kill it. I had kept quiet for a while but, that was the last straw. I told him to put a roof on his pen if it didn't already have one, and that he should think twice about killing any lion because if I found out that he had I'd turn him in. Mr. Savage gave me a scathing look as he sped out of my driveway. I hope he cooled off and

31

thought about how to remedy his situation. It would be important if he ever wanted to have more guinea hens.

From my discussion with the man's foreman, the pen was six feet tall. A cat can jump in or weasel its way around until it finds a way over the top. Hookah may have been too large to successfully get into the pen. If there were a small opening at the top between the roof and the pen walls, Ben could have easily gotten in. Hookah and Ben shared this love for a guinea dinner, and perhaps she was teaching him bad habits. They took two or three each time they would visit or so I was told by the foreman. It seems they made several trips to the pen, until there were only a couple of guinea's left. As it turns out, while baby was bringing home dinner, the second party's dogs were facing a death sentence because it was supposedly their fault. If anyone else could see how well fed these dogs are there would be no question in mind that they could not achieve the elaborate theft. Physically for them, it is impossible. In my experience, if dogs had been steeling birds, there would have been feathers all over the dog owner's yard. There were no feathers anywhere, except for the few lying next to the pen.

As the holidays grew closer, I did not see any more lion tracks or hear anything about missing guinea hens. The days had grown even colder and cloudier. During the mornings, I would look out the window to see if she were perhaps making her way down the side of the ravine into the creek bed. Other times I would walk out to the pond to see if Hookah were still taking her motherhood breaks, but there was no sign of her. She had taken her one remaining cub to a higher elevation to finish raising him where more food was available.

Chapter VI
Hookah

Fog usually plagues the Central Valley after the first few days of rain in the fall or early winter. While the lower foothill region may be hidden from view, the tops of foothills and mountains are usually clear and warm. This means that prey species' are more abundant at higher elevations in the fall and winter because the valley is under a blanket of clouds. The lioness Hookah prefers the rocky outcroppings in the winter for shelter, and food source availability. This is especially so when she has cubs. In zone 2, a wild goat herd was incorporated into the food chain for that habitat. Along with the remaining deer from the summer, and pigs that can be found at the base of the mountain, there is enough large prey to keep a lioness with cubs fed while she's there.

Almost one year later in November, a friend asked Tracy if we would take him and a guest on a hike up the mountain. Tracy was able to get permission to go ahead with the hike, so onward we went. My mare and I have a deal; we ride a couple of times a year. If I propose to her that we up the activity she looks at me as if to say, "This is not in my dossier," but, this day seemed fine with my big, fat, cupcake mare. Yes, I prefer riding a grouchy mare up that mountain than to walk. In zone 2 the elevation at the highest point is 1,720 feet. The views from the top of this mountain are incredible. The lake and surrounding cattle pastures are in proximity from the east side of the mountain, allowing a lion easy access to the valley floor. While riding up the switch back road, cupcake turned around and took a look at the valley below with wide eyes. She was probably thinking about going back home! It is quite a climb. I had packed an infrared camera, and one with a flash to see if I could get pictures of the wildlife on the mountain.

When we got to the last bend on the trail before going around to the other side of the mountain I noticed a familiar sight. A large cat had left scat on the dirt road. This I knew was the beginning of marked territory. As we climbed higher passing through perfect lion habitat, I noticed a second scat on the road. We had ridden through where the lion considered her front porch to be. The morning air was crisp and the view was awe striking. I can see how a person wouldn't realized how high up the top of the mountain is until you are there. I thought to myself, I wonder if there is any scent of cat urine around. This is probably not what most people would wonder but, I am always looking for signs of this magnificent creature. I turned my head to the side and gave the air two sniffs. My mare's ears perked up, she lifted her head, and then she too gave the air a couple of sniffs. I had to smile as I was thinking to myself, I wonder if I gave it away or if she already knew something was up. I think she already knew but didn't give it much thought until I called it to her attention. After all, how many people sniff the air for cat urine! I was checking to see if there were any mating pairs in the area.

After climbing up the other side of the mountain we split off from our son and guests so that I could place a few cameras. This did not please me. I told them to stay together. The scat that I had seen earlier had appeared to be older with some decay, so I really didn't think that we would see anything. Nevertheless, my statement to my son and guests was clear. As my husband waited patiently, I found a couple of spots on top of rocks that the camera focus would be clear from the tall grasses that sway and get in the way of the camera lens. Another obstacle of placing cameras was cattle. Since there were cattle on the mountain, I had to make sure to place them high enough not to be knocked over by a cow scratching its head on a rock.

On our way to meet up with our son and guests, we called them to see where they were. Unbeknownst to me at the time, they were having an experience of their own. They had found the flat spot on top of the mountain to break for lunch, and wait for us. Meanwhile one of our guests had heard something in a brushy rock pile. As he got closer to the rock pile he thought he heard growling. My son was a bit surprised but knew what it was. Our guest promptly threw a large rock into the brush and the growling stopped. While looking at my son, he tried to reassure himself by repeating, "That's not a mountain lion… it couldn't be a mountain lion." With my experiences, your heart and stomach know what that sound is before your mind does. Yes! It was a mountain lion. Luckily, it was the young lion Ben that hadn't been near people. All he wanted to do was hide to avoid any altercation.

While looking for our second party, we called them again. On the phone, the new arrangement was to meet back down at the bottom of the mountain. Now, the best thing to do for everyone was to go back down the same way we all came up. It is never a forgiving hike down the steepest part of the mountain, especially when your depth perception is deceived by tall grasses and hidden rocks. Unfortunately, this is the route they chose to take. A few bumps and thuds later, they had gotten through the worst part and made it back to the bottom in one piece.

Tracy and I went back down the mountain the same way we'd come up. The valley was clear of morning fog and the sun shone bright. What a beautiful day it was for a ride. Cupcake the mare was happy to be going home. Upon entering the area where the old scat was located I wondered if Hookah was there. My mare didn't seem to be bothered by anything and I was admiring the view. Some movement caught the corner of my eye; I looked up the mountain to my right and saw birds. They had all flown up into the air. I know that if birds are disturbed by another animal they fly away. Then to my surprise I saw a large animal on top of a rock. When I realized what and who it was, I knew I was looking at a mature mountain lion. This was not just any lion; it was the lioness, "Hookah." She had been watching us come down the trail and into her territory again, as it was clearly marked. While we rode closer to her first boundary mark, she decided to follow us. She was even with my right shoulder and about one hundred and twenty yards away at the most. She sat down briefly to watch us. After we'd traveled about fifteen yards or so, I watched her as she jumped from one rock to a much taller rock.

The increase in elevation would promote a better visual advantage for her. When the lioness began her jump with the seemingly effortless, powerful thrust of energy, her long body stretched out in mid-air as if she were flying. Upon Hookah's landing on top of the rock, all of her muscle definition shown in the afternoon sunlight. The form of that cat was beautiful. Her coat was shiny and there was no mistaking that "J" curve tail! Every move she made was in perfect silence. Hookah was curious to see where we were going. Again, she sat silently watching us. I could see her face. Her downward gaze was enchanting and her stature was regal. Unfortunately, the moment was soon interrupted by the wildlife I was riding.

My mare had seen her too; Cupcake cranked her head all the way up and to the side. If she could have, she probably would have spun it around in circles. So from then, all the way down the mountain, Cupcake worried. If a bird flew out above the rocks, it must have been part cat.

Hookah, had taken her one remaining cub up the mountain to finish raising him in an area that she considered to be safe with enough prey to support them both. She did a good job raising him and preparing him to go out on his own. Soon, Ben would leave the safety of the secluded mountain top to find his own home range. I have found that overlapping home ranges for mountain lions is common, but hierarchy seems to determine who inhabits the better habitat.

Meanwhile, my husband who was riding in front of me never saw a thing. As he and his well-behaved gelding went ting, ting, tinkling, from the sound of his spurs down the mountain in their blithe way, my mare was trying not to climb out of her own skin! I asked him if perhaps he might pick up the pace a bit. He replied, "Were fine." I could see Hookah watching us all the way down the mountain. I thought better of telling Tracy and attracting more attention to ourselves, so I just sucked it up and shook my head. Then the thought crossed my mind that my blond ponytail blowing in the breeze might have looked like a big, feathery cat toy to her. This bothered me.

On the way down the mountain I watched the behavior of the cattle. They too, were somewhat on edge. I had thought that it was in part because of our presence but given the lioness watching from the top of the mountain I could understand their anxiety. Their heads would not only turn to watch us but, would then turn back to watch up the mountain. I never saw her again that day. I wondered if she'd followed us at all. Perhaps she had made her way down the mountain along the rocks, just close enough to loose elevation and yet feel like she was keeping her anonymity. If this is what she had done, she kept a good distance from the cattle on the mid portion of the mountain. None of them decided to break and run, they knew she was there.

There are different types of cattle to watch for. The kind that are raised on a smaller ranch, that don't get real upset about anything. They know when dinner time is and come in on their own. Then there are the others, the cows that are raised on huge ranches that seldom see a human. Those are the kind that act like Henny Penny and will do anything to get away from you. These are the cows that I don't look directly at because their eyes get so big, and most of the time they're gone in a flash. A cow like this gives off the sense that it's going to explode with a huge burst of energy, and no one knows in which direction. Then there are the cows that will chase you down, as my older son found out one day. But, that's another story.

I have found with four different cattle herds that; cows and a lioness will share a mountain and work around each other. In the mornings the cows will graze the opposite side of the mountain from where the cat travels through. Later, the cows move over to a different area of the mountain because the cat has passed through and they are free to move over to the other side. I have seen where the cows graze right around the rocks where a lioness has had a young cub during the day. Knowing that the cub and mother lion are gone in late afternoon to early evening, the cows graze that spot and then move on to where they're supposed to be by nightfall. They have a system.

Our horses and mule have the same system. Usually they have no worries about where they go out in their pasture but, forty acres becomes rather small when one of the cats is on its way through. The horses and mule move at a hurried pace to the corrals just below the house. They turn and watch as the big cat travels through from the bottom to the upper pasture, then across the road. They stay up close to the house at night sometimes. This allows the mountain lion to hunt the pastures for rodents or coyotes. The next morning, they go back out. Sometimes, if it's Hookah passing through, they just move down to the other end of the acreage. The way they behave differently with each lion is interesting. The dogs will creen with a male lion where they do not when Hookah comes through. The two subadults only receive barking and then silence. When the crickets in the yard stop chirping, I know Hookah or the subadults are on the bottom hillside near the yard.

Meanwhile, when we met up with our son and guests, Tracy rode along talking with his friend. I asked my son how it went on the mountain. He replied, "Mom, you're never going to believe this," and he told me the story. At the bottom of the mountain when we were about to load up our horses and go home, I asked our guest if he saw anything up there. He looked at me with a puzzled look on his face and said something like, "Did you know there was a mountain lion up there?" I remember my reply. I told him, "No, I had no idea (until I saw her scat)." Then I reminded him that I told them all to stay together. I thought his emotion when he asked me that question was both funny and slightly absurd. It was kind of like it was my fault that wildlife was on the mountain when he wanted to go up there. I'm sorry; I forgot to notify all living creatures to vacate the premises for our special use. I was thankful that Hookah showed us the decency that she did. After all, how many times has she been to our house? This was the end of a good day.

The next day, as I looked out the bedroom window at Cupcake, she was lying down in the grass, so comfortably. She was really tired from the day before. I wondered what she could be thinking about. She was probably thinking about going on strike the next time I ask her to take me up that mountain. She's a great horse and I love her personality, but for no words out of her, she is quite the sarcastic one.

I started thinking about what I had learned about people from the day before. I was disappointed. There are people that really do enjoy the outdoors but, need to keep in mind that wildlife does exist in the country. If they had preferred nice terrain, and squirrels, they should have gone to a park. Oh, maybe that's it. They were out for the scenery and forgot about wild animals. I'm just glad there were no rabid raccoons for them to run across. Then what would've happened? While thinking about the lion cub's reactions to their presence, he had not acted in a confrontational manor. He remained in hiding. These people were lucky because essentially they put themselves and my son in harm's way by making him feel as though he'd been backed into a corner. Had he been older, he may have come out of hiding to make people completely aware that they were in the wrong place. This may have led to an attack. Instead, as the two other people followed the example of my son, and backed out of the area, the lion did not show his presence at all.

They were all lucky that Hookah was on the other end of the mountain. Ben stayed in hiding. He merely gave a warning out of fear to the intruders. The low grumble was to make them aware that they were to close for his comfort. When my son and I spoke about the sound the lion made, he said that the grumble did not sound angry. They had to figure out where the sound was coming from because it wasn't very loud. There were three growls before the older guest threw the rock into the hidden den. Then he stopped, there was no more growling. I worried that the person had hurt the lion cub and thought, "Now, was that really necessary?" This is another good example of someone acting out of fear.

Later, I was relieved and happy to see that the youngster was fine while on his way out into the world. Perhaps the rock throwing individual acted out of fear but, if it had been me, I would have just left the area. This chance meeting between the lion cub and three people may have taught Ben to stay quiet, and as hidden as possible when a human approaches. I hope that he will leave people to their own business and stay out of site in the future.

A few weeks later in December, Tracy and I rode back up the mountain to retrieve my cameras. There were no signs of any lions, the cows were content, and the afternoon weather was clear with a light breeze blowing over the grasses. We ate our lunch sitting on a rock overlooking the valley. In the back of my mind I had to wonder…is she still here? I found myself listening for any snap of a twig or weed. I looked everywhere on the horizon in all directions for her presence. I was hoping to see a hint of her again, but if she were the smart kitty I believe she is, she was either watching from a distant rock or out birding somewhere with her cub.

My camera pictures were amusing. I got a picture of a goat, lightning over the mountains, weather changes during the days that had passed and a silhouette of a cats face. Now, research is fun but it is still science. I check everything twice and sometimes more. There are two kinds of wild cats in this area, a bobcat and a mountain lion. I checked the picture against another profile picture of a juvenile lion, it matched. Because the photo was taken in the hours of darkness, the picture was black and white. I inverted the generic profile pictures color to black and white…and it matched. I checked for hair tufts on the side of the face and at the top of the ears like bobcats have, there were none. My thought is that the flash of that particular camera drew in the curious juvenile and I caught a picture of him. In this case it was a young felid that lacked life experience.

The other infrared camera was undisturbed with only scenery pictures. The problem I have found with cameras that flash is that the pictures are not really of the best quality. However, they do draw in the more curious wildlife. The mature mountain lions are smarter and avoid the area of coverage including the cameras flash burst. Infrared cameras work best because they have no flash and, I have had more success with them. I have literally scanned through thousands of pictures and videos, depending on the weather and surrounding wildlife, there is usually more than two species' in pictures. When checking photos, I make sure that I look at the surrounding background area also to confirm wildlife presence.

I have heard people say that when the line between people and wildlife is crossed, that's when wildlife gets dangerous. While that is partially true, it is man who is capable of making the decisions on how to deal with the controversy between him and mountain lions, and it should not always result in the lion's death. For some people the realization that they exist comes when seeing one. This makes people fearful and they start thinking that the possibility exists that all resident cats are out to get them. On the contrary, the majority of the time mountain lions are never seen. I have heard that there are people that have studied these animals all of their lives and have yet to see one in its own natural habitat. Management plans for mountain lions are important but, I think it's really the other way around. We need better management plans for people in terms of consequences for shooting one.

Mountain Lions know when a human is near, before a person realizes a lion's presence. They are so well camouflaged with just one dominant color that corresponds to their habitat that they are sometimes really hard to see in the shade. I can pick them up in a camera lens but the shadows make the pictures come out really dark most of the time. I have also been in situations where I knew I was being tracked but, I couldn't see the lion. I quickly learned that since I don't carry a gun, I either cannot go alone or I need to stay close to the truck.

It's no secret that Mountain lions can travel great distances because Sneaker travels a route that I only know parts of. He makes his rounds, and then comes back at a certain time of year. Other lions may be young and blatant about making their presence known when traveling by our home but, their presence is short lived. From my research on a young lion that hunts throughout the creek bottom, he only calls for a max of fifteen minutes and then he moves to a different spot somewhere up or down the dry creek. The difference between Sneaker and the young lion is that Sneaker stays out of close proximity to people. They may see him from a distance. Young lions are too curious. They often times come closer than they should (Ben). Hookah is very careful about where she goes primarily because of her cubs. I have noticed that her home range while she has cubs is small compared to Sneaker's. I think she's happy here. I would be disappointed to see someone decide to fear the fact that Hookah lives here. She was born here and she too raises her young among us. She shows her true colors whenever someone sees her. She acknowledges people with a look and continues on her way. I have seen this many times. It would be nice if people could just do the same for her.

Nature does move in a circle. If wildlife gets out of hand, it's probably because there are fewer predators of a prey species'. Unlike Hookah, she and others like her have no natural predators down this low in elevation, accept for man. Hookah may be a mountain lion, a large carnivore, but she has a role to fulfill in a rural setting. She is primarily a rodent hunter and without her and others like her to keep the rodent populations down, more pesticides will be needed to do her job. While it is true that if one keystone species' is removed from the equation, another species' will take over. This trade off still upsets the role of nature. There is a large biodiversity to support the mountain lions in their ranges. Having a wide range of species' in such a small geographical area shows a well-rounded ecosystem. This small basin surrounded by foothills brings people from other towns to bird watch, photograph the fox, beaver families, and other wildlife.

Thought: If a mountain lion in captivity eats about five pounds of meat a day and doesn't travel like a wild one, how many bobcats would it take to replace one mountain lion? With an increase in bobcat population, what would be the issue with small stock in pens that are not fully enclosed? Would this lead to an increase in small stock loss? Which species' would become more numerous, the bobcat or the coyote? What problems would the influx of coyotes cause with spring and fall calves, as compared to a resident cat that primarily eats rodents? Would the new loss ratios attributed to bobcats and or coyotes be higher in comparison than that of one or two resident lions?

Growing up as a child, the coyote populations were thick. There were many nights when the sound of their howling was right outside my window. Every once in a while my mom and I could hear a couple of them run through the driveway near our bedroom windows. As time went on their populations began to decrease. The cycle of a healthy ecosystem begins from the top down. I love it when there is a resident lion in the area.

Tracy and I were overlooking the valley from the top of Campbell Mountain. When the fog burns off, this is what a mountain lion sees.

Chapter VII
Ben

Subadult- January 22nd, 2011. I had been at the store, and when I got home found my son's dog at the end of the driveway to meet me. I thought this was really odd because this dog is pretty passive and doesn't mind being on a cable for a while. When Tracy went out to see what had happened with the dog and the cable, we observed large cat tracks in the driveway. The tracks came in the front gate and went all the way down the driveway to the back of the property, and then back again. This was a brave cat because there were four dogs in very close proximity to one another. The 17 month old cub wasn't experienced enough to do a lot of damage. With all of the barking, he probably figured out that he had gotten himself into more of a situation than he could handle. Apparently the young male had been here and spooked the dog.

It wasn't until a few days later on January 27th, 2010, when we actually saw him. It was just after 2:00 p.m. I'd been on the computer when Tracy walked in the door and told me to come look outside. When I looked towards the dirt road, I saw a yellow colored cat at a quick jog making his way down towards the olive trees. He was looking everywhere to see what was around him as he traveled. Judging by the height of the fence wires alongside him, he was about three feet tall. After he disappeared I decided to drive over to the fence line and take a look at his tracks. This youngster had some big feet. Following his tracks down to the olive grove, I looked for his entrance into the tall grasses that have taken over. Sure enough, he was hiding in the irrigation runoff slough under the cover of tall reeds, olive, and cottonwood trees. The trees and grasses are so thick that it was impossible for me to safely get out of the truck to see if I could find him. At least I know he had water and a safe place to rest and hunt before he moved on. I knew he was still in the reeds at the time because I'd circled around on the dirt road. There were only his tracks leading into the area. His mother must have taken him there to hunt ducks and other prey when he was little.

Young cubs use all of the tools that their mother taught them to survive. I have seen this when cubs become juveniles, and disperse. They travel to the same locations for which they became familiar, when with their mother's. Looking at the travel patterns of humans and lions, I see that like teenagers beginning to drive, they take the same routes their parents have. Then when a comfort level is reached, exploring new areas and finding new routes to get to new destinations becomes predictable. I think it could be the same with a young lion's dispersal. They travel the routes from their childhood looking for a niche and when they don't find it where they used to feel at home, they move on.

January 30th, 2011. The subadult lion (Ben) seems to have left the area. Sometime in February, I had heard from a Police Officer that the helicopter had spotted him with infrared in the process of doing their work. While checking out the area from above, the helicopter had found him instead. Ben had gotten stuck across the river because the water had risen. The excess rain water, or foothill runoff, has to be released from the dam because the snow pack will eventually melt and have nowhere to be stored.

April 2012- It was a little past 8:00 p.m., I was a few minutes late getting to the BMX track to pick up Bryan. Of course he didn't mind. Those boys would sleep there if they could. Since we were in the truck, I asked him if he wanted to run to Starbucks. He knew we were headed that way anyway and said, "Vanilla Bean?" Just after we turned onto a main road into town, my head lights clearly caught Ben's body flying across the road. I hadn't seen him in a while but, I knew it was him. I think his feet only touched the road twice while he was crossing. His body was fully extended in motion. When he reached the other side in what seemed like a split second, he went down the embankment into the slough, up the side of the hill and, into the stone fruit trees. All I could say was, "Awesome, Bryan did you see him?" Bryan thought that was pretty cool, too. Wow, now I know where he crosses into the Kings River.

It was a few weeks later when I was at the college speaking with an advisor when she told me that a mountain lion had been seen. I asked where the sighting was and what happened. Julie said that some students had been doing some work for a class down on the farm and had caught a glimpse of the lion's head peeking out just above the grasses at them. They said he took off when he knew he'd been spotted. I told her I was glad to hear that, and I knew who it was. I told her it was Ben, how old he was, and his story. I had also heard that some people walking had seen him in late afternoon hunting in the grasses, as well. I was told that he never bothered them, and they just passed by. He probably associates people with rock throwing! Maybe this is a good thing. No one saw him anymore after those two instances.

Young Adult- In June, Ben came back for his first seasonal range change since he'd disbursed at the age of 17 months. Ben is now three years and two months old. On a sunny afternoon around 3:30 or 4:00 in the afternoon, my dad, and Bryan were on their way to town for a burger. Ben was crossing the road just as my dad reached the top of the hill. My Dad stopped his truck and waited. The cat walked in front of him and sat down on the side of the road. My dad said, "He just sat there looking at me." He was a dark reddish brown with black spots, and he had those big yellow eyes. My dad and Bryan were so excited to sit there and watch a lion that I had talked about for so long. Dad said, "We sat there for the longest time and he just looked at me, he wasn't worried about me at all. He looked at me like he needed some help." I told my dad he didn't need any help except maybe to find a girlfriend. I explained to him that it was his first seasonal range change since he'd left, and that he was now breeding age. Dad laughed and said he didn't think he could help him there.

My mom told me that she was doing yard work and my dad came home bursting through the back gate to tell her he'd seen the lion. Mom said he was so excited about it that he was grinning from ear to ear! Bryan cracks me up. He watches me do this all the time, so he's not that excited. Bryan has gotten a lot of driving practice by following me down every dirt road in the study zones. As long as he has food while accompanying me, he's great! His appetite is close in comparison to the lion's. He can eat two Porterhouse steaks at a time. My dad is great at keeping the tri-tip burritos flowing for Bryan.

On another night, the dogs started to howl which then turned to creening. This went on until 10:30 p.m. They only do this when a lion is present because they're afraid. It also happened like a chain of events. First the dogs down the creek began creening. Then it becomes silent as the cat gets closer. The cat made his way into the area. Our dogs began creening while the mule complained. It's a warning that told other animals where the cat was. It took Ben about twenty minutes to get here or close to our house.

Apparently, he wasn't in a big hurry. It takes him about ten or fifteen minutes to travel one third of a mile through the Ag land where homes having dogs are sporadically located. Weaving through these places, Ben followed the same route as when he left here.

The next morning, I got a call from my mom. She told me that their dogs were creening as well. My parents live about 2 miles away to the east. Ben had gone sightseeing. The canal was bank full so he crossed over the bridge that separates one habitat location from another. I wondered if he used the bridge or not so I looked for his tracks, and found them crossing from the opposite side of the road to the bridge. Of course I didn't tell the neighbor's this, it would've upset the balance of things. For the most part, he's pretty safe as long as people stay away from him. There's a reason for the times during the day or night that he travels, and it's not only to hunt. No one has moved into the path of his travel with more dogs that would divert his routes from his destinations.

After our dogs showed concern for Ben being in the area, I took a rake out to the established fixed points for his frequency of travel. When I rake the ground, I can find out what part of the day he's coming through. This is a great technique that helps to identify wildlife in the area. Perhaps Ben not finding a girlfriend while completing his rout was the reason his stay was so short.

After dinner when it was about time to call it a day, the dog's creening began again. It only lasted for a moment or two. Ben was closer than the dogs thought. It got so quiet outside that you could hear a pin drop. In the past, it was quiet before Hookah made her hunting sounds, too. I woke up during the early morning hours to hearing the dog's for another few seconds. Then Ben made his intension's clear, he was hunting. I could hear him whistle for his dinner out in the back. When he began, he sounded like a kitten. Then he turned to a rabbit scream. Yes! He was trying to kill two birds with one stone. If he sounds like a rabbit either a rabbit will poke its head out for a look or he will get really lucky and get a bigger dinner of coyote. I hope he eats a coyote. His calling stopped after about twenty minutes. Early in the morning, Ben started intermittently calling again. From my IR pictures, this is about the time the coyotes start making their rounds. His hunting calls stopped around 6:00 a.m. I think his hunting was interrupted by the spray rigs in the citrus. The workers start earlier than normal when spraying pesticides.

No Date - It's clear to me that Ben likes to eat birds, and rodents, although he prefers animals of the *Canis* persuasion. I knew he was here because the dogs creened off and on for half an hour. Then I realized they had ticked off the cat! Ben was now down in the bottom near our house mimicking our creening dogs. They took turns doing this until I yelled out the window for all of them to, "Shut-Up!" The dogs felt better and I whipped out the IR night vision. I looked everywhere for that cat. I can see a wide perimeter around our house, from inside. I couldn't see Ben's shiny red eye glow anywhere. After that I couldn't sleep and turned on the lights in the kitchen. All seven dogs and the wolf know what happens when they wake me up! The shells are in the kitchen drawer. A couple of warning shots usually ensures a good night sleep for me. Poor things…between Ben the lion and the kitchen light flipping on, they probably wondered which way the night would end up. I knew he was still out there.

No Date - Last night and the night before, there were short periods of creening but no more than three at a time. Once on Saturday in late afternoon around 4:30 p.m., the dogs told me that Ben was traveling through the upper pasture. This is where he crosses the road to go squirrel hunting around the old, overgrown, pomegranate tree. Ben has learned to stay away from the dogs, and the house. Lions are so smart. I wish people could catch onto things as quickly!

No Date- It was early evening. I took my camera out and set it up on the tripod. The sun was going down so it was going to be iffy with the light aspect. I waited a while and then there he was. Ben was taking the dirt road around the creek bottom to cross over down into the pasture. He saw me in the yard and stopped to take a look at me. It was short lived as he kept moving at a quick pace to get to cover. Funny thing, the dogs made a little noise but, I think because I was out there and he was only passing through during the day light hours, they knew he wasn't serious about his hunting. I got a picture of him but, his shape is distorted from the shade of the orange trees on the hill. It's crazy how parts of them can be almost invisible in the open. When the horses saw Ben for the first time crossing down into the pasture it scared them and they ran like race horses to the safety of the corals. After a few

47

times of him crossing where he does and at the point in time that he chooses, the horses would automatically just come up to the corals or move over to the end of the pasture giving him a wide berth for his crossing, in the evening. At night, they knew it was better to be at the corals and they spent the nights down there. Their system works.

No Date- Tonight the horses were at the other end of the pasture awaiting Ben's passing through. It was about the same time of afternoon/early evening. He didn't come from the same direction. I had clear vision across the pasture with a larger camera lens. Again, the shadows saved him from me getting a great picture. He came off of the dirt road, down into the pasture, up across the table top and back down near the pond. As he was slowly beginning to become visible again, he passed under a eucalyptus tree. I took the picture. I can see the partial shape of his body, long tail, ears, and the white part of his muzzle. There is a hanging tree branch dead center in the middle of his body! This was another spoiled picture, not to mention that the lighting didn't help. This time when he crossed he went into the orange trees. Ben hunts squirrels at the fence line and bunnies in the orange groves. This is one reason I do not take walks out there. When I set a camera, I drive to the location. It will be interesting to see what time he comes back to hunt tonight.

No Date- Sound carries along the creek bottom. I can here dogs to the east of our home. He's still traveling through here to spend time in zone 1A. This zone is small unfarmable land connected to zone 1. It is separated by a widely traveled road, and canal that runs May through the end of July. He will eventually end up behind my parents' house. I'll call them in the morning to see if he showed up.

No Date- My mom said she heard Ben hunting last night. I wondered how long it would take him to find his way over there. He must've gone by way of the lake. I know he crossed the bridge of the canal. She first heard the dogs behind their house. She went out onto the back porch for a while so she could hear the sounds he makes while hunting. The dogs got quiet, the closer he got. Then she heard it, the faint distress call of a rabbit. Before we built the house that sits behind my parents, there were a lot of coyotes and feral dogs out there. So many, that we had to bring our cattle home. Ben remembered this from hunting with his mother as a cub. He set out to see if things were the same. Things are not the same. The people that bought the house have sheep. Thankfully they are smart, having good fences, a roof on the pen, and Great Pyrenees dogs in with the sheep. If only all people with stock would do this, there might be fewer complaints about lions eating small stock. My mom heard Ben call for about 15 minutes off and on. Then the night fell silent again.

The fact that Ben hunts coyotes is interesting to me not because they are larger prey but, because it means that before his dispersal his mother may have had him hunting the coyotes with her. This is highly possible. It was good training for him. It's also of interest because of his location before dispersal. The back of the lake had a lot of coyotes that roamed the citrus and vineyards that separate the lake from the mountain. The coyote population does seem to be significantly smaller when we have a cat in the area. The nights are also free of their yipping. There is always a group or two of coyotes that roam the area but, they are either fewer or more cautious. I definitely know that they are fewer when the cats leave.

Out of nowhere mom heard one shot from a neighbor to the west of her on the main road. Then three more shots rang out. They were from a different neighbor a half a mile at most, down the same road to the north. So far I haven't heard anything from anyone else about the lion or his location. Firing warning shots is acceptable. I do this sometimes when the dogs are worried about something outside. It's meant as a deterrent for wild animals, and it brings peace back to my sleeping hours. It may only be teaching a mountain lion to keep distance from the house but, that's OK. I would like to think that they won't get into trouble anywhere else but, as for now I think they only associate the warning shots with certain locations.

Ben had left zone 1 at the house to travel across the bridge into zones 1A, 2 and 3. However, on this date I found that he has expanded his territory to encompass more farm land of stone fruit to the south of all zones. In order for him to get to the location where I found his tracks he had to walk down a paved road, and cross a busy intersection either early in the morning or late at night. He was happy to get off the pavement onto the soft shoulder of the planted field and traveled down the road to see some of the neighbors' pigs. Ben didn't bother the pigs but, he did take a

49

look. He didn't circle around to a different bridge; he came back almost the same way he came to cross the road again traveling north to zone 1A. It's possible that he will circle around into habitat near water to hunt. Cougars follow pigs down every slough.

A few mornings later- Ben came down off the mountain early this morning. It took him less time to come down the mountain that it takes a rider on horseback to climb it. He took the long way around our house, through the citrus trees, and down into the creek bottom. From there, his rout would lead him back to the slough, then into the Kings River bottom. There is close to seven miles between zone 1 and the Kings River before he can reach the crop land on the other side.

One evening, I was at the College playing tennis with my son. The bright light at the courts seemed to make the surrounding darkness blacker. It was my sons turn to serve. As he drew his recquet up into the air, my attention was drawn to a familiar sound. Off in the distance came the warning call of dog's creening. My heart leapt with joy! Ben had made it home. He was across the river. Needless to say, I missed Bryan's serve. I was trying to triangulate in my mind, how far off he was from the location that the dog's creening came from. I wondered how far he would go from there. It took Ben ten days to get back into extended home range. The acreage across the river is largely agricultural farming of stone fruit and row crops. The Kings River ends in Tulare. The property he has access to is beautiful. Ben has the safety of the meandering, overgrown river banks to roam or catch a nap during the day. He bothers no one.

Chapter VIII
Sneaker

February 12th, 2011. I posted two cameras near a holding pond where I'd found lion scat and tracks. The pond is at the bottom of a foothill. The pictures did not reveal a mountain lion but there were a few nice pictures of red winged blackbirds. A raccoon got caught by surprise, and two coyotes tried to slink by. The most frustrating footage to look at came from camera one. There were almost two thousand pictures of one very busy squirrel trying to protect his home. When I had aimed the camera, I should've made sure there were no squirrel holes in front of it.

While collecting my cameras I found out that Sneaker had been in the area. As an added bonus, Hookah was with him. The two lions were together on February 8, 2010. Hookah had gone into estrous because Ben had dispersed. Now it was time for her to mate again. By going back to January 27th when we last saw Ben, it could have taken a week or more for Hookah and Sneaker to find each other. The time between Ben's dispersal and the two lions being seen together is nineteen days. Subtract one week for a grace period for them to find each other. This puts the date at February 2nd, give a couple of days.

I don't know the system that they had but, I think it was due to areas of scat left for communication. It is a possibility that Sneaker knew her home range bordered along his, since they both grew up here. With respect to the adult lions seen together, all track and sign, and the date that I stopped finding two sets of tracks from either lion together, the mating time lends its self to the estimation of maybe two weeks. I say maybe because during the last few days of tracking, their tracks were still in the same vicinity, yet not close together like a couple.

While tracking the pair, I found that they were hunting bullfrogs along the slough of the canal. I found their tracks in the mud along with a sow and her piglets. From the size and length that the lion and pig tracks were apart, I can estimate that the lion was chasing the sow. There were drag marks in the clay mud, off into the grasses. This confirms Sneaker caught the sow and perhaps he and Hookah shared a meal. I saw Hookah's tracks in the chase as well. It is conceivable that she caught her own dinner of piglet. Nevertheless, they were within close proximity of one another and

51

all lion tracks headed for the grassy field. This field was not fenced. It had been a grove of over grown olive trees a couple of years ago. The trees were pushed out to make room for more productive citrus or grape farming. The old piles of trees hidden within the shoulder tall grasses made the best hiding place. If I were a lion, I would hide or take naps there.

During the time that I had found tracks of both lions being together, Hookahs tracks were never far from Sneakers. At the end of what I estimate to be a little more than two weeks, approximately seventeen days, from the 2nd of February to the 17th of February, the tracks started to become farther apart. Soon, I found no tracks from Hookah and few tracks from Sneaker. He'd remained in the area for close to two more days. From the decrease in tracks from both lions in the places that I had been finding them, it appears that Hookah, and Sneaker parted company between the 17th and the 19th of February. The last tracks of Sneaker I found were heading out of the area to the east. This direction leads across a bridge, to a foothill chain across a semi-busy road to higher elevations.

These mountain lions are very familiar with crossing bridges. When I check to see if Sneaker is in the area, I often check one bridge on a remote section of the canal. He remarks his territory when he crosses that bridge each time he's in the area.

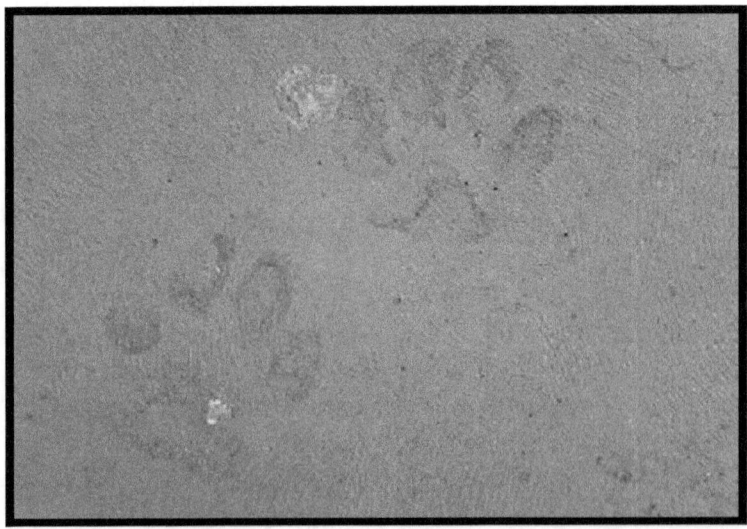

Sneaker left his tracks on the concrete. These tracks are great for identification.

Casts made from Sneaker's tracks when he was 8 years old.

March 2nd, 2011. I left the house at 6:00 a.m. to go out and pick up a cast that I had left from the night before. I prefer not to be out at dark because I don't want to interrupt his hunting hours if he's here. The tracks and castings were all by the old olive grove. The soil moisture was high and the wet plaster mix took more time to dry than I had allowed for. While I was there, I took five more casts because I thought that I didn't have a full set of Sneakers paw prints. I only had three. When I looked at the pictures of his tracks made on the concrete, I realized that I did have the fourth one. The right rear print of the foot pad is different than the other three pads that are intact. It appears that he had an old injury that took off half of the outside pad.

Walking further along a slough I saw more tracks from Sneaker. He was hunting. His tracks followed skunk, pigs, and a squirrel tracks. Along the canal runoff ditch, there was aquatic life of bullfrogs, and ducks. Sneaker had gone back to his solitary life style while Hookah resumed her life of walking on grass and debris to avoid the presence of any other male. She was now expecting a new litter of cubs.

March 31st, 2011. A friend stopped by today and told me that Sneaker had been there, and he had been washing muddy lion tracks off of the porch. I took pictures of them and discovered some differences and with the casts of his tracks, I can identify him anywhere.

January 22nd, 2012. Human's actions can be a little scary sometimes. While I was out doing a little research, I was completely surprised. I have never had an animal scare me like this two legged did. Between the end of January and beginning of February, I was tracking in a remote spot close to where I'd found a calf carcass in two pieces. I wanted to track all around where I'd found the carcass to make sure I hadn't missed the tracks of a lion.

Out of the peace and serenity came alarming sounds that I couldn't place. They weren't the sounds of the cow that had escaped from the pasture. Though she wanted to be in my back pocket, she stayed under an old tree near the pasture gate. From the way the cow was acting, and the fact that she was out, it most likely was the calf's mother. No, this was an alarming, fast paced shaking sound accompanied by pounding footsteps. My first thought was that a young bull had gotten out and was in a hurry to get down the hill through the citrus groves. As I listened closely it didn't sound like the thundering steps of a heavy four legged. I stood still, waiting to see what was moving in my direction. Katie was following behind me in the truck. As the sounds grew louder, and closer, I could see a head of dark hair emerging over the tops of the grape vines. A man was running down off of the mountain through a row of vines and trees. He was running so fast that it sounded like he was heading straight toward me. Then I saw the gun he had slung over his shoulder. It was black and looked like an assault rifle but, it was a pellet gun.

That was the shaking noise that I'd heard, with the pounding thuds of his feet. There was nothing I could do in this situation but wait for events to play out. The man passed in front of me, then slowed to a walk. He acted like I wasn't even there as he passed. At this point, I had no idea what to do so I just simply said, "Hi, how are you?" He looked back, nodded his head once, and kept going. I looked at Katie who was sitting in the truck, Aw struck! All we could do was laugh at each other's reactions to the surprise situation.

The man that had come running down the hillside avenue perplexed me. I know that while there are cultural issues with the way some people in the United States survive, I think we're all under the same government. The possibility of someone living off the land is not widely accepted out here.

There are some places here that people do not frequently travel; it makes this area excellent habitat for wildlife. The outskirts of town are not the wild outback. I think it's important to let the wildlife survive.

With all of my tracking around the general area of the carcass, I did not find any tracks of a lion in the area. There is a well-used cat door in the shrubbery. There were no entrance or exit tracks from any felids. I found many tracks of coyotes. The calf was in a pasture. I venture to question what could have happened with the aid of coyotes. The last thing I thought about when he came off the hill was that maybe something was chasing him. I'd looked around after he passed by me and saw nothing. Perhaps I had been the one to alarm him with my unexpected presence. I have no definite answers.

April of 2012. I received information that Sneaker was allegedly shot. The person who mentioned this said he wasn't sure if he was killed or not. The incident supposedly happened in the middle of January or February. Part of that information was regarding a calf being eaten. Then soon after, Barbara was seen. If predation by a lion were a factor, she'd be the most likely candidate. Although I was not made aware of the location of predation by this person, I had found an issue concerning predation myself. The calf carcass I had found looked like it would have been a calf weighing about 450 pounds alive. The cows missing calf was about this weight, and very healthy. She had no problem walking up close to a human. A calf this size, and healthy, wouldn't be a target to a mountain lion but, would be to a human. To date, there has been no sign of Sneaker. It would be a shame if Sneaker had been killed due to someone's desire to "live off the land." I hope this wasn't the case.

Chapter IX
A New Lioness

March 3rd, 2011. Today, I took my mom with me to see if we could find new tracks, but there was nothing new in any of the locations from the prior week. There was one last place to look. The bridge at the end of the mountain is usually my last stop to see if there are any lions traveling from the neighboring foothills. I found four sets of tracks. One set was off to the side of the road. They were smaller than a half dollar. The other two sets of tracks belonged to Sneaker. But there was a new set that I hadn't seen before. They were much bigger than a bobcat, not the same shape, and smaller than Hookah's tracks.

This set, along with baby tracks were coming from the orange trees across the road. This was at the edge of zone 3, the border of my research project. Across the road to the east lies a foothill chain that I'm not able to access yet. The bridge is a common drinking place and a crossing area onto the foothill in zone 3. The area is secluded and people often pull over here, have a small fire, and drink beer. You can see the remnants of old fires, beer cans, and various other things. There was an old pile of peanut shells on the ground. I guess Sneaker didn't think much of peanuts because that's where he chose to leave his scat, right on top of the pile.

Note- Since I question Hookah's age and there could have been a litter from her two years prior to her having Ben and Snuggles, could I be looking at her first cub whom is now an adult? Would it have been a single female born to her, earlier? The age discrepancy between her being six years old to eight years old would be solved. If this was her first cub then she would be about eight years old and the lion I'm looking at would be close to three and a half years old. It would make sense because from what I've seen, the females don't venture far from where they were raised. Maybe I'm looking at a set of tracks from a new female. If I am, she has a cub. The tracks are maybe about a week old. I hope it rains again this week for more fresh tracks.

Some time had passed before I saw any movement from a resident cat. It wasn't until the middle of April that I found evidence that Hookah was back in our neighborhood. I woke up one morning early to a rabbit scream down in the bottom pasture.

It was either Hookah making the noise or the rabbit. There was a poof of fur where the rabbit met its demise. This was part of Hookah's profile. The only thing I needed to find next were the feathers of a bird to show it was definitely her. After taking a walk out on the neighboring range land I found what I was looking for, a poof of bird feathers. Upon finding two more poofs, I had to laugh to myself because the ground nesting birds don't have a chance with this cat. She's like a kid at Disneyland, only she's bouncing from nest to nest. Perhaps this is another reason she likes to relax on the pond bank. She can see where all of the nesting sites are located. This is one smart kitty!

April 15th, 2011. My husband and I stayed up a little later in the evening to finish watching a movie. When the movie was over, Tracy opened the bedroom window before we went to sleep. I asked him if he really thought this was a good idea. He told me not to worry, that if a lion were around it wasn't going to come in the window. Later, an irritating sound woke me up at about 2:30 a.m. I woke up mad because I thought he was snoring again. This had gone on long enough! I sat up in bed and looked at him. Odd, I thought, it sounded just like him except when I looked at him his mouth wasn't open, and he wasn't snoring. The noise was still going on. Perplexed about this, I got up to look out the window. I nudged Tracy's foot. He wouldn't wake up so I smacked his foot repeatedly! This made him jump and he was definitely awake now. I said, "Tracy, listen!" Tracy was noisy and said, "What, listen to what?" I said, "Shhh…listen to that sound outside the window." Tracy asked what it was. I told him, it's Hookah, she's hunting our cats! So, he sat up and said, "Oh," and then went back to sleep. I was not impressed with his lack of excitement.

Meanwhile, I carefully stood back from the window because it sounded like she was right underneath it. As I quietly moved closer to the window in the dark, I could see that she wasn't at the window like I'd thought. No, she was three feet away from the window around the corner of the house, hiding in the shadows. This is where I heard her vocalization for hunting domestic cats. The sound was a low toned broken m-e-o-w, with a click at the end of every sound she made. She was almost enunciating these sounds into words. It was so awesome to be standing there, just out of her range of sight and to hear how she was hunting her prey. I had always heard that lions imitate other sounds, but this

57

was the first time I had experienced this. I had clear visibility out of our bedroom window into the yard. Our dogs knew what this sound was and all seven of them including the wolf, were silent. The ducks were in their nest, and the rabbits were hiding at the backs of their cages. Out of 16 cats, there was not one that could be seen. Hookah wasn't getting a response from my cats so she became more forceful with her calling. She grew louder and more intent.

Safely out of her sight, separated from the huntress by the corner of our house and a window screen, I could feel my heart beat harder, and adrenaline rush. I felt my stomach tense up with each deep guttural sound she made, as though I were her prey. My skin felt different. It was like all of my senses were heightened. It was a good opportunity for me to see what it was like to be the hunted, while out at night during the prime hunting hours of a mountain lion. With this experience I could see what I needed to do to control my physical emotions during this type of encounter. Had this been an actual encounter at night, there would be several things that I could do to avoid it. Being aware of my surroundings when in known lion territory at night, not go anywhere in the dark alone, or diffuse the situation. Take the cat off guard, if I were not the hunted. This is what occurred next. I said in a loud stern voice, "Hookah, what are you doing?" You could have heard a pin drop in the still night air. There was total silence, I never heard her leave. I waited to see if she would come out of the shadows, but she didn't. What a great experience this was. To be so close to the lioness that I had watched for so long when she'd not even known I was there. Hookah was pregnant at that time. She had come back to den in the empty pond bottom, under the roots of an old fallen tree. The den is surrounded by tall grasses that exclude it from vision. I knew she was here because on the way home from town in the late afternoons, there were days I could see her lying in the afternoon sun on the side of the pond bank.

Looking back, I think I made a mistake. Perhaps I should not have said anything to her, because three nights later she came back. She was completely silent about stealing one of my cats. She learned not to call them because it didn't work in her favor. She also didn't want to draw the unwanted attention from a human. Since then, Hookah has been back to the house twice for two more cats. With my interference, her porch hunting skills improved. She learned to be quiet about it. A thought just crossed my mind about regular house cats. When a truck backfires, and a cat doesn't expect it, they involuntarily fly three feet up in the air. I wonder how high Hookah would fly if I'd banged pots and pans together. It's a good thing there's no window above our bed. I would have seen how high Hookah and Tracy would both fly up in the air! Yes, I really do want a window over our bed.

April 20th, 2011. It was Monday morning of spring break. I took my daughter with me to go look for a place to put out a wildlife camera. I had asked a farmer if I could set up a camera on his property prior to my visit. I had explained to him what I was really looking for in hopes that he wouldn't have a problem with it, and even though he was a bit surprised, he gave me permission.

I had been looking for a large male that weighed close to three hundred pounds, and thought that maybe I might find some sign of him in this location. Sneaker used to stretch out on the big flat rock at the top of the hill for a nap in the sun.

I drove up the mountain and around the citrus trees to park my pickup. My daughter was on the phone, so I left her in the truck while I walked partially up the mountain. I wasn't sure my truck would make it up the mowed road because the grass was wet. It was a beautiful morning around 9:00 a.m., and the air was cool. As I walked up the hill looking around to see if there were any signs of lion scat, I approached a large rock mound to my right. While I was walking by, I caught a glimpse of something moving on top. As I got closer to the base of the rocks, out of nowhere walks a lioness. I stopped and looked at her in disbelief of what I was actually seeing. She too stopped to look at me with the same surprised look. I was about thirty yards away from her. I had frozen in my tracks. All I could do was watch her. In the first few seconds after our eyes met, she swished her tail a couple of times, then sat down. A brief moment later, after the initial shock of seeing one another had worn off; we both just looked at each other. At that moment, I knew what I was seeing but couldn't find it within myself to be afraid.

This was clearly a moment where we were learning about one another. She appeared to be amused by my presence. Even though I knew what she was, I found myself checking her body and facial features, one glance at a time. While she was in the sitting position, she swished her tail up and down two or three times.

Then she blinked her eyes and showed me that she had become more relaxed. The pupils of her eyes were still fixed on mine while she evaluated my intentions, and what she was going to do next. While she sat, her ears were forward. Her body was not rigid with fear, just relaxed. I know my breathing had become shallow for a moment but, my heart was not racing like I thought it might in this precarious position. She was studying me for a potential meal.

While looking at this lioness, I couldn't believe how beautiful she was. Her face was dainty, her under eyes, nose and muzzle were beautifully outlined in black. She was a very petite cat. Her ears were not small and round. They were tall and narrow because she was still young. Her face was small making her ears look a little large for her head. Briefly glancing at her ears, I looked to make sure there were no tufts at the tops of them like bobcats have, then turning my focus back to her pupils.

When I started checking out her features with briefly displaced eye movements, she did the same with me. While looking for the tufts on her ears and seeing they were absent, I looked at the color of her ears to make sure there was no white spot. No white spot was present. They were perfectly black but with the morning sun shining down on them, they looked a little brown. I looked at the sides of her cheeks to make sure there were no poofs of hair, like bobcats have. Her facial markings were crisp. She had two black stripes from the corners of both eyes that stretched back almost to the bottoms of her ears. The two black arm bands were present, not yet faded. Her eyes were big, bright, and yellow. Her nose was orange but, not as broad across as a male's. She was a soft creamy yellow with large faint yellow spots on her back. As the young lioness sat in front of me, I could see her white chest, and fluffy underbelly. Her body was sort of thin and lanky. She had thin forelegs with big paws. She sat up on top of a rock in a wide crevasse. I would estimate her sitting height close to three and a half feet.

I found out first hand, that it is an absolute must to keep eye contact with these creatures. The whole time we were in each other's presence, we read the body language of one another. Physical actions and reactions can make for either a relaxed visit or one that won't turn out so well. False moves must be avoided.

While I looked into her eyes I saw that her pupils were wide. To me, this means that I saw no sign of aggression. I suppose mine were wide as well due to the surprise of seeing her. After I was sure that I was in front of a mountain lioness, I just stood there...looking at her. I didn't want to leave, who knew if I would get this chance again. The opportunity to stand so close to a lioness that showed no actions of being anything other than, "At home," was the experience of a lifetime.

The moment soon grew old. As I went to bid her farewell I made the mistake of a partial smile. Her pupils quickly grew skinny, her facial features changed in a split second as she licked her lips exposing each of her sharp, shiny white canines as a message to me. When I realized what I had done, I stopped immediately. Then her pupils became wide again as to only give me a warning. I took it to say, if you show me your teeth, I'll show you mine, and mine are sharper. Then her eyes and facial features softened. Her overall demeanor went back to the relaxed happy cat that she was a few seconds before. With point taken I decided to have two more seconds with her. Then I nodded my head to her, started to carefully back up, keeping my footing small and even paced. As I got farther away I turned to the side and was careful to keep her in my peripheral vision for as long as I could. The grasses were three to four feet in some areas. Even though she showed no signs of aggression, she is a big kitty and I know how cats love to follow a person. My domestic cats stalk me all over the yard, and to the mail box. She never left nor made the effort to move at all while I backed up. She was so sophisticated, sitting tall on her rock overlooking me while I made a timely exit.

Later, I realized that I had felt as though I had been in the presence of a queen and was to act accordingly with the proper etiquette that a courtier from a story book, would display. Before this visit with a mountain lion, I had no idea what that really was. It dawned on me that turning my back on a queen would show disregard. As I walked down the hill, I looked back a little at a time to make sure I didn't have a four legged shadow. This visit with a beautiful mountain lioness in her own habitat was more educational than anything out of a book. I was just lucky that I was not only afforded this chance meeting but, that it went well too.

By the time I had almost reached the truck, feeling like I had just been promoted into the world of greatness, I saw that my daughter was still on the phone. She had missed everything! I needed a way to identify this lioness. I used to number the cats but it became easier just to name them because they are not all from the same lineage, or so I thought. The name for this lioness is Suka. It means, "Fast," in Inuit. I chose an Inuit name because they inhabit such remote arctic and sub-arctic regions. I can't think of anything else more solitary than that.

After the rare encounter with Suka, I went home to get a camera stand because there were no fence posts or trees to mount the camera to. When I returned to the area, I drove up the hill and I was surprised again. A small felid scampered across the rocks where I had seen Suka. At first glance, it was dark in color with yellow spots. Yes! She has a baby! Suka had allowed her cub to come out of the rocky crevasse to soak up some sunshine, despite the earlier visit we had. My daughter was finally off of the phone.

Kate helped me set up the camera exactly where I had been standing when I first saw Suka. We waited and watched but, saw nothing more of them. I thought it would be at least marginally safe. When I retrieved my camera, there she was. I knew she couldn't contain her curiosity. After we had left by just a few minutes, down the hill she came to inspect what we had left for her. The picture shows Suka in the tall grasses that are blowing in the breeze. I can see the outline of her face and there is no missing those ears! The outline of her back and tail, are present as well. I find that it is necessary to always invert any picture that appears to be a cat like creature to see if a long "J" curve tail is present, and it was. I had of course already seen her tail when she was in the standing position but for the sake of the picture I had to check it again. There is a strong possibility that she could be Hookah's first cub. From what I've seen with these lionesses, they seem to only have one cub in the beginning. Or at least one cub lives out of two. I have also seen that a males color is sometimes darker as a cub. Then they change. Nevertheless, the cubs I've seen were from Sneaker. He was reddish brown.

The information I gained from this rare type of visit is as follows: Suka's fuzzy baby's eyes were open but, his tail was not long like I thought a cub's tail would be. He was scampering quite well. If I estimate his age from his size and take his activity into

62

account, I would estimate his age to be about three months old. Then I count backward three months from April, and Henry would have been born sometime in January of 2011. The barely visible yellow spots on Suka's back told me that Suka was young. The arm bands on a mountain lion fade out when they get older, like Hookah, but Suka's arm bands were still very black. I think that with the uncertainty of a few months that I can respectively age her close to three and a half years old. Now, given the month and date that I saw them, it is likely that Suka was pregnant with Henry in November or December of 2010. This estimate allows for one or two weeks. Suka would have been born between March and May, of 2008.

To end the notes, there were multiple scats from this lioness found all around the rocky outcroppings. The scat consisted of small bones, and fur from most likely birds, and cottontails. It is safe to say her diet consists of small rodents. Her approximate weight is estimated at one hundred to one hundred and twenty pounds.

Note on Behavior: When Suka and I saw each other, the meeting was between us. Her cub was not out on the rocks with her. When Katie and I came back, she would have seen us driving up the mountain had she been napping on the boulders. When we had gotten close to the rocks and saw the cub, we did not see Suka. The cub ran from the rock where he was sunning himself, back down into a hole between boulders. I think this is where his mother would put him while she went out to hunt. Though we didn't see her, she may have been hunting in the tall grasses that surrounded the rock mounds.

Henry stayed in his designated area. I thought that a cub would stay down in the rocks. Instead, Henry showed the discipline to stay close to them but also, the initiative to venture out into an area that was not beneath the rocks. He chose a place where he felt secure. Some of my more feral domestic cats will have their kittens under a large berry bush. The kittens don't come out to play, they stay under the bush. As they get older, they will venture out on their own but, only go so far away from the security of the bush unless their mother's call them. I see a similarity here. Both species' have a radius around their dens that they will venture to. The feral kittens stay with their mother after she allows them to come out of the bush, to travel the yard with her. They do not venture very far from her. Later, they become braver, and the mommy cat's let them have a little more wiggle room. They put them in the tree house. The kittens wait until their moms are out of sight. Then and they come down to play on the ground under the tree. I have seen one mommy cat in particular; give her kittens, "The look." The kitten that is misbehaving backs up into the tree zone, and acts like nothing happened. This tells me that as a mountain lion cub gets older he may stay close to or in proximity of his mother but, he expands his little territory from her. She is either watching him or out hunting but, she knows where he will be.

Chapter X
A Visit from Hookah

April 21st, 2011. Currently I am still looking for Hookah's den. My question for researching Hookah for her second litter is: Will Hookah return to den in the same place she did last time? The answer was partially yes. Hookah did return to spend some afternoons on the pond bank but, something was different. I only saw her for about a week or so.

Usually, a lot of tall green grass is at the bottom of the basin, creating a nice play yard for her cubs. When Hookah had her first littler, she spent quite a bit of time in and out of this area beforehand. It was time that she be in an area where she planned to den. Her cubs would be due soon. I walked over to look at the pond and to my surprise the grass was brown, and dry. Hookah is a picky cat or should I say, very selective.

There were tire tracks all through the bottom like someone had driven a small tractor around in circles. I could see the tire tracks of the small spray rig that went over the edge of the bank and into the bottom. One of the workers had sprayed weed killer in the pond bottom and killed the little ecosystem making her prior denning spot undesirable to her.

Hookah had decided not to use this place after all. I started looking for places that she might be. There is an old tree that overhangs the creek bank at the back of the property. I can't see underneath the bows due to dense foliage. There are old coyote dens dug out of the hill side near the base of the tree. During the prior fall season, the leaves had dropped and the caves were visible. One of the old coyote den holes is large enough that if she chose to use it, she could have. I know that mountain lions are not cave dwellers but the bluff, and tree do provide shelter, and safety. I will check out the area again in the fall after she's had her cubs. If this were the area she'd chosen, then there should be some signs of old scat nearby. In the past, I had put a camera out there to find a family of raccoons not far from the tree, down at the creek. There were prey animals there for her. The grass was green everywhere else. My wolf hybrid had found fresh lion scat at the fence line.

April 23rd, 2011. Today, I'm taking my mom with me to change the SD cards in the cameras. Sometimes, there is no activity at all that pertains to lions. This gets old but it makes those hard to come by photos of tracks just that much better. I have heard people say there are too many lions. Others say there are no lions or low numbers of them. I think the prey plays a huge role in where they center their home ranges.

It is my thought that the lions push downward in elevation, was due to the increase of human population inhabiting the lion's natural foothill habitat. At this time, mountain lions are still seeking the last tiny bit of treed, rocky hill tops left to them. The foothills that border the valley floor have been the lion's last resort. Their wandering nature has led them down into agricultural farmlands. Within their search for small inhabitable areas that are not agriculturally suitable, they are encountering more people and their anonymity continues to be threatened. The three major lions within my study zones are finding the last of the natural springs, ephemeral creeks, and ridge tops to inhabit. They have had to resort to man-made ponds, ditches and canals for drinking water, adapt to farm work and traffic, along with the occasional risk of accidentally running across mountain cyclists, joggers, and walkers. They are only seeking what is natural for them while trying to avoid humans at the same time.

The resting locations Hookah may choose depend on where the sun hits the mountain tops in early morning. She and her growing cubs, spend time on the tops of the foothills. They nap on boulders to soak up the sun in mid-morning. Later in the day she takes her cubs to rest under the canopy cover of trees that hang over tops of boulders. All the while, she has a perfect view of people down on the valley floor, coming and going from their homes, farm workers doing their daily jobs, the irrigation people on the canal banks, and the occasions when she sees me looking at her through a camera lens. In the heat of the summer, I find an increase in activity near valley riparian areas. The citrus groves that line the foothills and riparian habitat become hunting grounds for small pray in the late evenings, through the night, and early mornings. On long hot days when the sun doesn't go down until 9:00 at night, I have found that when the citrus is being irrigated through drip system, that she uses these groves more often. The water cooling the air in the evenings brings refreshment from the day, when it is even too hot to hunt near the lake bottom.

All of the lions prefer shaded areas, and can easily be overlooked while in the shadows. The field workers irrigate at night. Some of them have expressed worry for the possibility of running across one of them in the dark. I tell them not to worry. The sound of their ATV precedes them. The mountain lions don't want to be seen.

Note- association or recognition: In the beginning of the book, I wrote that a lion needs only to experience a disturbing situation with humans once to learn a particular behavior. This subject brings me to Hookah.

I drive my little white truck onto the canal banks to park. I then take out my camera, and scan the tops of foothills taking photographs in one continuous direction. Most of the time, I cannot see her the first time I shoot pictures. The second time around, I go back to the same starting point and there she is. Hookah lies on a giant boulder near a tree looking down at me; her cubs are usually close at hand. She has her favorite spots. I still wonder if her association is just with my small truck or if it's because of recognition when I wear my hair up in a clip that allows the ends of my blond hair to blow in the breeze. I wonder if this reminds her of the day she watched Tracy, and I ride down the trail through her domain. She sat above us on a rock looking down at me. The fact that I call her by name may mean that when she's in a given area, she may recognize the tone of my voice. She knows my tone of voice from when I spoke to her. It could also be that the first time we met; she heard my voice and saw me as well. Later, she may have recognized my truck because I use that specific truck the most when I track, and check fixed points. Maybe it's because the sun gets too warm on the other side of the mountain. The heat forces her to travel to her favorite spots near shade, and by that time I'm in her area looking for her. The partial issue with this is that I don't always go at a specific time or day. Perhaps it's coincidental, and the glare of my camera lens in the sun attracts her attention. I think Hookah does associate my truck with a ditch tenders truck and the glare from the lens catches their eyes. However it still doesn't account for when she's not in an area I am, and then shows up in a location where she can see me.

Cats are nosy. She doesn't make her presence obvious to me when I drive the white dually. In fact, I don't think I've ever gotten a picture of her when I drove the dually. I do have to think about the fact that she has seen me more than once over a number of years now, and she does pass through here every year. Over a period of time she has learned not only who some residents are but where they live. This tells me that she does not cross the line of being wild even though we (residents) have become familiar. When Sneaker would watch the vineyard workers tying vines and singing, he knew they were busy and didn't associate them with any physical danger. He left them alone while they worked. He just seemed to be amused by their activity. My observations show that these mountain lions are not trying to stalk or attack humans. They simply find us amusing. They want to keep their distance from all of us. I also find that people, and their behaviors or habits are the main reason for issues with mountain lions.

Hookah is a large cat, while Clause is almost her parallel at his young age of almost 13 months. I have a picture of the rock, and tree with them present, and one without. I had to do this for the memory of location. I may have some sightings of her while taking photos, but when I get home and download the pictures into the computer, they get even bigger. As I go through them, there are no photos of any lions until I pan back to where I started, and then there she is. Hookah is so curious about what I'm doing that she goes to a rock pile under a shady limb, lays there and watches me watch her. The truth of the matter is that sometimes she blends into the scenery so well that it's hard to see her.

The habitat in zones 1, 1A, 2, and 3 is suitable for longer periods of time because the climate is controlled by water. In places where there is less agriculture, the air is dryer. Where there is ground vegetation in these zones, there is more rodent prey. In the spring, tall grasses on hillsides make for good hunting during the early morning hours, when the area is mostly inaccessible to humans. This is especially true when Hookah has cubs. I've learned that Hookah is done hunting for the morning by 8:00 a.m. or so. Then it's nap time. She takes her cubs, heads for a hill top that overlooks the agricultural farm land, finds a rock that's shaded by a tree or two, takes a bath, washes her cubs a bit, and goes to sleep for a while. When it gets a little warmer on that rock, she takes her cubs to a different spot that has more shade cover. Along the way, she looks for a snack. Then, its nap time again. This happens all day or until she and her cubs need a bigger meal, a drink of water or it gets hot enough that they need to seek the shade of the orange groves to hunt small game. I noticed the same schedule with Suka, and Henry. Their solitary life styles are threatened, so what better place to go than where people will overlook them. I was told by another biologist that there are at least six to seven mountain lions within a one hundred mile radius. I think it depends not only on location but, their preferential diet, and number of lionesses with cubs.

May of 2012- The lioness, Hookah, concerns me. It's not that she tries to get caught in the open, but people continue to encroach on her habitat and interrupting her hunting hours. I'm not saying that people shouldn't ride their bikes on foothill cow trails, just that it would be nice to see people think about what an apex species' needs to live in this area. I was happy that these people got to have the experience of seeing Hookah. However, after they had seen her hunting during the evening hours not once but twice, it would've been better if they'd considered changing their route while she was there. I lost track of what was going on but, I hope that's what they did. It would be a shame if someone got hurt. Part of the blame would fall on the human for promoting a potentially dangerous situation. Ultimately, if someone had been attacked, she would pay the price.

I would like to see more serious penalties brought to individuals that are negligent in regards to the provocation of carnivores; specifically referring to livestock predation, and persons putting themselves, families, or friends, in a potentially harmful situation resulting in an altercation causing the taking of a mountain lion, with further penalties for causing the impending death of mountain lion cubs. As much as I would love to go post a camera where the people had seen Hookah, I would rather just be happy with the distant sightings of her I've had. Spring will come again. Her frequency of travel will increase in the lower elevations. Perhaps one day I will actually get a good photo of her but, that time is not now. Hookah and her cubs come first. Just because I see a marked or unmarked trail for that matter, does not necessarily mean that I should use it or take my family or friends for a walk in that area. Through what I have learned from the mountain lions that I've studied, they try very hard to stay away from humans. Even when they see one of us, they allow us to pass or they quietly pass by, without fail. It is more important in favor of the mountain lion to, not tempt fate. I feel that jogging trails should be monitored, be in populated areas, and people should jog in pairs or groups. The same goes for hikers. Awareness of our mountain lions is important and goes hand in hand with personal responsibility of educating one's self.

There are days that I go out to look for tracks. At times I know where resident cats are in certain areas. I choose to leave them to their daily rituals so that they may have peace. When I visit places that show frequency in travel, I sit in the open, and look around before getting out of the truck to go track. Hookah is almost invisible in the shade. When she's resting on a rocky hill top, she's hard to see if I'm not looking through a camera lens. If she and her cubs are in the vicinity, they see me. Sometimes I have the luxury of seeing them from a distance. Keeping a distance from a specific location where one of the cats may be, keeps them safer.

There are days when I've gone to research a specific zone, and when I get there, I look around, and listen. Even though there are large areas of land for them to roam, they are limited with so few places for them to find refuge during the day on flat land. I've gotten out to an area to go tracking for Hookah and her cubs to see if they are eating well but, just decided to go home. Sometimes it's better to be happy that they have a hideaway. I never go tracking through the orange groves. I stay on the outskirts when the weather is in the triple digits. The nocturnal hours of a lion or lioness can be quite busy as they look for a meal. If they do not find one, they hunt later in the mornings. I live in a unique location that affords me the opportunity to hear where the resident cats are in regards to hunting or just traveling.

An interesting fact of lion travel is that on the nights they are in the area but not in close proximity, the nights are silent. It's almost too quiet to sleep. This is especially true if Hookah is hunting the pastures. The dogs do not creen with her presence. They never have. The creening only happens when there is a male in the area. When Sneaker came to find Hookah, the dogs may have been a little on edge with his presence but, they never creened like they do when Ben is here. Hookah can silently walk up the embankment on the edge of our yard, and the dogs are silent. The gray lion passed through the pastures keeping his distance, and the dogs never said a word. Sneaker must have kept his distance as well because I really never heard much. They might have expressed a little discomfort but, not like they do with the young one. Ben hunts like a pro. Listening to him is amazing. He can reach such high pitches with his calling.

I've seen Ben come trotting down the dirt access road by the pasture in the early evenings. He knows where he's headed. When he reaches a certain point on that road he stops, and looks at the house. He watches to see what's happening over here before he continues on. I was out in the yard one evening and saw him while he sat and watched me. His coat was a bright shiny red in the late afternoon sun. When he was satisfied I wasn't going anywhere, he continued on his way with that bouncing kitty trot, down the dirt road. The dogs were not satisfied that I was outside this time. They let loose with their alarm calls. It didn't matter though because he knew they weren't coming after him.

Chapter XI
Mr. Inconspicuous

During the summer of 2010, Mr. Edwards was driving home from work. It was around four o'clock in the afternoon. He'd reached a busy intersection. While waiting to cross, he looked to the right and to his surprise; out of nowhere came a gray mountain lion. The big cat strode across the road. This is not a common occurrence here. Mr. Inconspicuous had begun crossing the major highway with traffic coming in both directions. Mr. Edwards thought he'd surely see the demise of the big cat. Just as the lion had reached the center of the road, a concrete truck driver slammed on his brakes in order to avoid hitting him. The long tailed cat had gotten safely to the other side of the road where he sauntered off into an empty pasture, and down into a dry canal system. He was lucky that day…people put the lion's safety first. The rout he was taking led him back behind the mountain where he had been seen before. There were two or three of the lions using this well-known mountain as a thoroughfare.

May 4th, 2011. The eye witness accounts of people that reside in the area are important sources of information. Tonight, while on a break from a science class, I heard a story from a classmate that spends time out here with family. He told me of a gray mountain lion that he and his father-in-law had seen two years prior. This lion was traveling with a mate. They saw the pair crossing open land together. It was in late afternoon on a cloudy day. In the small valley where they were seen, the thunderheads are purplish-blue, and the dry golden brown grasses bend to the direction of the breezes. He said it was amazing to see especially because he thought he was lucky to even see one. The pair came from the north east heading towards the south end of the valley. It was then the men lost sight of them.

The man that shared his experience with me had also mentioned that an individual had trespassed on the private property soon after this story got out. He believed that this person was there with his dogs to hunt the lion. It was conveyed to me that when the hunter's presence was noted, he was told to leave and never come back. Though shots may have been heard, a carcass was not found. It is possible that the lion was out of reach. On a different afternoon, I was checking the pond behind our home for track or sign. I ran across Mr. Smith who lives in our neighborhood. We got to talking about Hookah and he was laughing at the name I gave her. He told me his wife was alarmed on an early morning while drinking her coffee. She had asked him to come to the porch to look outside. When he did, he saw the gray mountain lion casually walking across the lower pasture heading towards the pond. Mr. Smith's wife was surprised to see him so close to the house, even though the other two lions used the same access. Nevertheless, the gray lion continued on his way and never gave his penned up dogs a second look. They didn't see him anymore that day.

Chapter XII
Surprise!

May 8th, 2011. I took my mother with me to set out cameras today. I try to take someone with me when I go for the sake of safety. Not everyone has field research that can eat them. Besides, its quality time spent. I had explained to my mother that when I find fresh tracks and set out a camera, I do not bend down without someone watching my blind side. Being aware of my surroundings and having backup is equally important.

Sometimes a bird will give a cat away with its squawking. Birds flying away abruptly can be a clue that something is out there other than the wind. When I told my mother what I needed her to do, she said something to the effect of, "Oh don't worry I'll be watching." Since that's my mother, I didn't think twice about turning my focus to other things. I found a great place to set a camera. There were fresh tracks on the soft clay soil. The camera stand was in a great spot and ready to mount the camera to. It was slightly hidden by two orange trees. The day was beautiful. It was a really quiet, warm day with just a slight breeze. My mother was in the truck with "A book." This was not what I had in mind but, I figured she'd be able to do both. I was concentrating on where the laser aim was hitting the ground to make sure I would get the best coverage. Knowing that I was well within Suka's center of home range, and in a place that she frequented, I paid close attention to everything. I know how silent she can be. I listened to every sound no matter how faint.

I was almost finished setting the camera when I heard a crunching sound. My body froze in position, again. I thought, "Oh no, I came at the wrong time." Listening further I could tell that the sounds of crackling orange tree leaves were close. As I scanned the ground beneath the limbs of the trees, I looked for the paws of a mountain lion or perhaps Suka's face peering through the leaves at me. There was nothing in sight. I slowly rose up to stand as tall as I could. I used my peripheral vision while I slowly turned to see where the noise was coming from. With anticipation to see some type of animal, I turned all the way around and saw…nothing. Nothing accept my totally relaxed mother, sitting in the truck immersed in slowly, unwrapping her candy bar!

All I could do was watch her. When I felt enough relief, it seemed easier to giggle to myself about what had just happened. She had no idea what I had just been through. That my heart had skipped a few more beats. I finished setting the camera, and got back into the truck. I just looked at her. She was still sitting there, happy that we were spending time together. I ate a candy bar with her, and we went home. I'd thought about keeping it to myself but, that just didn't happen. Though I will not record what we said, we had a good laugh about it later.

Once, I had someone ask me if doing this type of research had ever made me feel like the hair on the back of my neck had stood up. The answer I gave that person was, no. But, that was before this outing with my mother. Despite my small internal panic attack, I still take her with me sometimes. It's amazing how a candy bar wrapper being opened slowly can sound so much like crackling, dried up, orange tree leaves. After that visit she became more aware of what we were actually doing. She also got really good at finding the girls tracks.

May 11th, 2011. While Henry was still small, Suka's home range was approximately 7.19 miles around her favorite resting spot. Her grassy yard, fence line and her area for hunting are located in two different types of habitat. About a month later when Henry was bigger, Suka took him with her more. She had found her way to another place suitable for cover, hunting, with plenty of water, and seclusion from people. They would travel between these two places more frequently than others. Suka also used to take Henry to fields with tall grasses for hunting pigs. Soon after, she was taking him to locations where she could hunt deer. They didn't spend much time on Granite Mountain anymore.

In the beginning, Sneaker occupied Granite Mountain. A part of the mountain property had gone up for sale, and was vacant. He remained for a while after new owners had set up their farming operation. After Sneaker took off to go see some girl cats, a different girl moved in. A few weeks ago, he came back only to find he now shares his mountain. Suka's scat marks her area, and Sneaker's scat marks his. The difference in size of scat is how I know which ones are his. It also helps to differentiate between the two scats to have seen his for a while before Suka moved in. Tracks will be easier to find again after the rain. How funny, they used fence lines, fence corners, and sloughs to mark their territories. I guess they have land marks too.

Later that day, the wolf and I went for a walk. She found more of Hookah's scat near the electric fence at the creek. I know she chose to den in a different spot but, I still can't find it. Even though she's not at the same address I know she's still in the area. I've seen her on the pond bank at least four times in the past two weeks. Her beautiful cream colored coat has turned darker with a reddish brown tint and she has really filled out. She looked pretty happy out there. Maybe she's had her cubs and, they're sleeping. The coyotes have been almost nonexistent in this area which tells me she's here at night, too. She can stay as long as she likes.

May 15th, 2011. Tracy went with me to retrieve two cameras today. One camera had a silhouette picture of a bobcat staring into the camera. It was funny to see because the camera was mounted on the top of a "T" post. The bobcat had to stand up on its hind legs to see into the camera face. There were no more pictures of Suka. Tracy and I walked up the grassy road to look for her. We found nothing, but when we neared the last big rock almost at the top of the hill, I asked, "Tracy, do you see them?" Tracy said, "See what?" Then when he turned around he saw that he had walked by the two bobcats that were still staring at him. We were both surprised at how well they blended into the rocks. They looked pretty comfortable, not worried about us at all. I figured we'd better go so we walked back down the mountain leaving the bobcats to bask in the sun at their leisure.

A family of bobcats uses Granite Mountain for sunning themselves.

June 4th, 2011. While driving by the lake today, I stopped and talked to my neighbor, Mary. She told me that two days prior she'd seen a lion crossing the street from the orange trees, to go down to the lake side. I asked her if the animal she saw had a long or short tail. She said the cat had a very long tail. That's how she knew it was a lion. I knew exactly which one it was, too. It was Hookah! I hadn't seen or heard anything of her in about six weeks, and three days. The description of her coat color matched what I'd seen. Mary was closer to Hookah when she'd seen her, than when I'd seen her on the pond bank. I asked if Hookah had seen her watching as she crossed the road. Mary told me that she and her dog were standing in the middle of the road watching Hookah. It didn't bother her in the slightest. As Hookah gave her a look as she sauntered across the road. I was so happy that my neighbor had seen her. The big cat stays to herself and out of trouble. By this time Hookah had her cubs. They were about one month old. The big cat was coming from the orange trees because there is cover to hide her while she walks in the late daylight hours. I still wonder where she'd found her new denning spot. Perhaps she found one near water where there is plenty of aquatic life. I'd spent a few minutes tracking Hookah through and around the orange trees to see which way she'd go. Her walking patterns varied. Sometimes she'd take the third row to the end then cross the road. Other times she'd travel down from the middle. The direction she came from each time was the same.

The eye witness account from Mary was informative. Her demeanor made their encounter comfortable. She is a small framed, quiet person. Mary's behavior during her sighting showed no aggression, and she made no noise. She was also completely visible as she watched the big cat crossing the road. The lion was aware of her and her old dog's presence. The dog didn't bark at the lion, only watched from her mistresses' side. Their body language didn't attract any attention from Hookah. Due to Hookah being a resident here since birth, she is used to the presence of people. She wasn't worried about Mary or her dog. She was seeking cover in the late afternoon. Hookah had only acknowledged their presence while she went about her business. Hookah did not run away, and she wasn't worried about being seen. Day time encounters with mountain lions are different than at night. Mountain lions hunt at night and behave differently than in the day time. Mary has since moved but, while she lived at the residence for two years, she had cats and a dog that spent their time outdoors. The wildlife is abundant with gray fox, many ducks, raccoon, opossum, and a Great Horned Owl that built her nest in a Eucalyptus tree. This resident had co-existed in close daily proximity with a mountain lion that used her backyard for water, cover, and hunting. There were no confrontations. Another nearby neighbor, "Jane," walks her dog along the same road with orange trees separating their homes. She has yet to see a mountain lion. There are approximately 275 yards between these two residential homes. There is a 271yard average distance between five residential homes located on the other side of the lake. This lioness has managed to travel the same walking paths and occupy the same lake property that cattle do and yet, there have been no losses due to this lioness. Karen, a neighbor two houses down from Jane, told me she was walking her daughters lamb when she found some very large tracks. They were Hookah's tracks. Neither she nor the other four neighbors had ever seen the lion, let alone lion tracks in their area before. Coexistence between humans and lions is not usually what one might think of when the word, "Mountain Lion," is used in the same sentence. This information goes to show that people are sometimes oblivious to the wildlife that coexists alongside them. In this case, Hookah was imprinted at an early age by humans, unlike Suka. Suka's behavior was more secretive though she had been watched by the pecan grove workers, verses Hookah. Hookah will periodically be seen but, only by accident. In the past when she has been seen, it was because she was trying to get to one of her travel locations by way of the citrus or nut groves. Her demeanor has always been business like even when she felt a familiarity with the human residents. This familiarity has posed no danger to humans as she keeps her distance.

Note: In June of 2011, I'd started a summer job with the U. S. Forest Service as a Wildlife Biologist Technician. The study of our resident cats had to be slightly put on hold. I did manage to keep track of their locations however, lost some time seeing where they'd gone, and how long they'd stayed during a week's time. Nevertheless, I was so happy that the job enabled me to take my tracking skills and apply them to looking for a different type of species'.

Chapter XIII
New Residents

July 17th, 2011.   Along the route that I take to go tracking in zone 2, lives a couple I have known for a very long time. Joe and Elena spend much of their time outside and see a lot of wildlife action in the evenings. When I stopped in to say, "Hi," this morning, they said they had seen a large bobcat. They told me I could set a camera up at the back of their property. So I did, the camera was facing the creek side with running water. There were plenty of tracks from opossum, raccoons, and coyotes. I even found some deer, and pig tracks. With thick green foliage, the location was perfect. This is where the bobcat would go under the fence to go down into the creek. The resident lions have a door like this, too. It's funny how wildlife does. If you were to look at the smallest creature on the animal chart, a bigger one follows it, and so on. It is this way with predators and prey. For example: a raccoon takes a walk, and the bobcat follows the raccoon, the coyote follows the bobcat and or the raccoon, and the mountain lion follows them all. Their tracks give me the whole story. There's usually a squirrel involved, somewhere.

After I set cameras, I drove to a place where I had seen Hookah's tracks. I was not disappointed. I found big Hookah paw prints. Behind hers I found the tracks of three little cubs. Hookah had three babies! They would be about two months old. I didn't know they would venture with their mother at that age. Perhaps my timing is incorrect but from the approximate date she parted from Sneaker, this would be very close. If my timing is off it would only be by one week. The babies' tracks were close to the size of a half dollar. I could tell Hookah's cubs tracks apart from Henry's not only because of the number of cubs but the size of their tracks. The tracks of the little yellow cat were bigger than a half dollar coin.

I followed the tracks Hookah and her cubs had left, across the road and through the fence. Hookah left a hair sample on the barbed wire. Maybe she was in a hurry while moving them to a safer place. They were too young to go hunting with her. Looking back, I think of Suka and her baby boy when he was small. After I had seen she and her cub, I knew that she had moved him as well. It is possible that they were just big enough to take them for short walks hiding them in the tall grasses while she went hunting. There are a lot of coyotes that would take advantage of the cubs as a meal so; I know she couldn't have been far. Coyotes are natural predators of lion cubs. The survival rate for cubs is difficult when there are too many, as Hookah found out when she lost one of her cubs to coyotes. This is the reason why, before a lioness dens, she clears the area she will den in of lower ranking predators.

July 24th, 2011. After finding that Hookah had taken the cubs into the lake bottom where there were plenty of birds, ducks, and other prey, I also found that she had been taking them with her more frequently. She changed their hiding spots often. Sometimes I would find only her tracks, and then other times they would all be together. Their age is close to three months now.

When I went back for the cameras at the neighbors, the pictures revealed a large, spotted bobcat. This cat really gets around; I've seen him before in a few other places. I printed the pictures for the land owners. People like to see the nocturnal wildlife that they miss out on while they're sleeping. They get just as excited as I do when they get the pictures. Besides, it's good for people to see that there is still natural beauty in the world.

To end my day, I went to look at another location of a fixed point. As I suspected, Hookah had already been there, too. I observed one set of lioness tracks with the three cubs' tracks. Hookah had been in pursuit of a small deer. The deer tracks skidded as they went back across the road into a field. The little cub's tracks were following behind mom, trying to keep up with her or hurry to her call. The time of the kill had been early in the morning as none of the tracks were there the evening before.

August 6th, 2011. There were pig tracks heading east, away from the orange trees. Suka was not far behind. She took down her meal. It probably weighed about 50 to 75 pounds. Her single cubs' tracks followed behind hers. It looked like he was in a hurry. The tracks tell the story of Suka hunting for dinner to feed her cub. I marked the waypoint where the skid marks stopped, and the drag marks began. Drag marks can be deceiving. I can't tell what's being dragged without seeing the tracks of what was caught. Pigs, and deer are both hoven but the shape of the pigs hooves are different. In wet soil a pigs tracks are easier to read than on hard pan.

I spoke with a man who works in the pecan grove where I track along the slough. I asked him if he'd seen a large yellow cat. The man said, "Yes," in an enthusiastic tone, and then he told me he had seen one baby with her. I knew it was Suka, and Henry. The food sources in that particular area are great for teaching cubs to hunt. There are a lot of bullfrogs and small rodents. I went back to tracking and found a small lion scat that was left where a prior scat by a mature lion was previously GPS marked. Then, to my delight I found perfect tracks of Suka's cub. His tracks are a little different in shape than the other three. He is close to seven and a half or eight months old now. I was able to get casts of them and a picture, too. It had rained. The mud hardened like a mold. The casts of his baby feet are so cute. He and his mommy were heading towards an open pasture that is surrounded by olive trees. This is the coolest place to escape the summer heat.

Last notes for the day: Suka has her route in an area that overlaps with Hookah's. The ground they travel, in some parts, is shared by them and it's hard to believe that they don't see each other. Both of the females share their habitat without seeming to have any problems. This is another reason I think Suka could be Hookah's first offspring. They pass through zone 2, sometimes only minutes apart. The moms are always looking around.

August 7th, 2011. I've done more tracking and GPS marking of sign. Every day there are new tracks that tell what the two families are doing.

August 11th, 2011. My work season ended and I'm looking forward to next season already. I love this job, I never get tired of field work, and when I find what I'm looking for it feels pretty good. At least I have my own project that keeps me busy with mountain lions at home.

August 12th, 2011. This afternoon, Tracy and I went out to the slough and I found fresh tracks from Suka, and Henry. They were leaving through the olive trees heading east. All of the cats are still using the bridges while marking territories with scat. I found something odd today. When we left the one section of small habitat, we drove past the back of the lake where there was sign left by a male lion. He sprayed a small tree. I could smell it as we drove by. We went back so I could look for tracks. Sure enough the tracks were Sneaker's. At this time both females have babies so there's no reason for him to be looking for a mate in this area, unless he is looking for Barbara Alice. She would be back from the river by now but, I don't know anyone who has seen her. I haven't found any of her tracks yet, but there is another place I need to look. Sneaker sired Hookah's babies and I'm pretty sure Henry is his too. Suka may have come from this area, and spanned out to the east, to find her niche. The area that she chose is also part of Sneaker's range, so the possibility of he being Henry's father are great.

Another neighbor just a few miles away told me of his shock one night while irrigating. He told me it was a couple of years prior to now. He was late getting home from work. The sun was almost down when he went out to turn off the irrigation to the citrus trees. When he bent down to turn off the water, something huge jumped out of the tree. It had jumped over his head and all he could see when he looked up was this long tail flying by his face. It was a large lion! The poor cougar had gotten a surprise because the man had spooked him. The man said that he never saw the entire cat because it was moving so fast to get away from him.

In the summer months the temperatures soar into the triple digits. I have also noticed that the three dens (only one in the study) on the foothills are all pointed towards the north. This could be coincidence but, perhaps it's to avoid the sun coming up in the east, or the rain that comes from the west, and possibly the summer sun that radiates downward, and from the south in the afternoons. Regarding irrigation; at the beginning of each row of citrus is a small standpipe low to the ground. Inside is a handle that turns the water on or off. When the irrigation water is turned on for the drip system; either the hose or the valve will leak. When the valve leaks it creates a pool of water hidden under the shade of a tree. This creates drinking places for wildlife. The water comes from the canal system. I have found mountain lion tracks in just this type of situation. This is the most accessible clear, cool water for drinking when it's too hot to walk a long distance to a watering hole.

Here is an example of the stand pipe for irrigation. The water has drained into the soil but when the valve is open the water pools, creating a fresh water drinking hole.

Photo of Henry's tracks made into casts.

Suka and Henry spent most of their mornings here in June through October of 2011. Hookah and her cubs would wander through a few yards away from where Suka and Henry were walking. Both families were in the same area, at the same time, on several occasions.

Chapter XIV
Hide and Seek

August 16th, 2011. Today, I tracked all morning, checked cameras and did some GPS work to delineate routes for certain lions. Sometimes, I have to admit that I feel like I'm following a lion that is somewhere up in a tree looking down at me, probably thinking that this is supposed to be the other way around. Who is stalking whom? I used to watch a cartoon about Dr. Livingston with the little man that wonders from place to place looking for him? Well, that's me. The only difference is that my research knows I'm there and just waits for me to get a little closer…so they can move again. At least that's what it feels like sometimes.

August 16th, 2011. This entry was written on August 27th. Later in the day, I had some errands to run with my two children. We were on our way to town when someone ran a stop sign. The driver of the small car hit my Ford F-550 on the front right fender, causing my wheels to lock up and subsequently we rolled three or four times, and took out four trees. There was nothing I could do to have avoided the situation; I never saw the oncoming vehicle until he was just a few yards from the side of my truck. This was a serious accident, and my children and I are still recovering. Tracking mountain lions and other wildlife is on the back burner for now.

August 29th, 2011. It's been almost two weeks; Tracy took me for a drive to get me out of the house. Sitting in a vehicle with broken vertebra is difficult. Standing, and moving is better for me. He took me out to the canal slough so I could see if there were any tracks from the girls, and their cubs. He and I walked down the trail a short distance when I decided to tease him a bit. I called out the cats name, "Here Suka, here kitty kitty," He just looked at me in that stern way that says, "Why are you tempting fate?" I asked him if there was something wrong. The pain killers I was on made me a little loopy and my sense of humor grew ornerier. I never thought that Suka would actually be there. We walked for another short distance so I could see if the beavers were out, and look for any sign of felids. The whole time I listened and watched for any sign of birds flying away, tracks, or scat. Then I saw the grasses moving down the hill side by the bank. I thought, "Oh Joy," we've got something moving! I said, "Look Tracy, she's over there," when really I expected it to be a squirrel or something. He did not find this amusing but appeared to be interested while glancing over at the grassy spot. I could not contain my orneriness, and laughed at him here and there when I wasn't listening for sounds. The weather was warm and the sky was a bit hazy, but there was a nice breeze that made the small habitat cooler near the running water. We had walked far enough for this trip. We turned around to go back to the truck. Have I said that I make it a habit to leave an area the same way I've come in? On the way back we heard nothing but the breeze. But, lying on the trail where we had just been a few minutes ago was a brand new fresh mountain lion scat complete with the curly-q at the end. The tell-tale color of green was there and, the smell that goes along with it. Once a person catches the scent of lion scat the first time, they never forget it. Not that I bend down to smell it because I don't! The pungent aroma fills the surrounding air. I looked all around where we'd been. I couldn't see her. It had been only minutes since we'd been standing there. She couldn't have been far. I wish I could've seen her again. It would have been nice. I got to thinking about this later, and she was probably in the culvert pipe right underneath me. It would be my luck, and her way of saying, "Tag, you're it"!

I have read in different sources that the scat is about nine inches long. This is true for what I have found. Sometimes the scat is sometimes just a bit larger around than a quarter. I think it depends on the age, and sex of the cat. But, on an average that is pretty close to what I have found. Needless to say, Tracy was ready to go home. I didn't feel really great physically, but I was on cloud nine mentally. I know I've said this a few times already but, I was really excited. I get that way when I'm closer to finding the kitty than I thought. And to think, Suka was there the whole time watching me look for her. Wow, how many people get to play hide and seek with a lioness that has no intentions accept satisfying her own curiosity? However, this is why a person should not go alone when looking for a large carnivore, or hanging out in their habitat. Tracy was not excited; he thought that at this time, we really should go. I on the other hand, dragged my feet a bit to see if I could see the cat that could apparently see me. Common sense did take over, and we left. But, in my mind on the way down the dusty avenue to the main road, I thought to myself, she was there!

August 31st, 2011. All of the cubs and their moms are present in their haunts. The days are beautiful. They have plenty of game to choose from.

September 1st, 2011. Today was the first day of hunting season. I am happy to see that all of the lions made it through the day. It was close to evening when Tracy and I drove to the fixed points to see how the wildlife had fared. I know where to look for the quickest results on who is where, so it was easy to find their tracks. Although they are nocturnal hunters for the most part, that is not always the case. Relief was my feeling when I could see that at this point the tracks were where they should have been, and so on for the next day as well. "The girls," sometimes I refer to the two lionesses as "the girls," usually travel the dirt access roads early in the mornings when it is still dark. It was probably a shock as well as an interruption, in their day to see, and hear people when no one should have been around at that time of the morning. If I were a lioness, the sound of vehicles, and their headlights flashing on the landscape at an unusual time of morning for traffic, would make me think twice about my daily activities. I would keep my children away from oddities like this and, under cover where they would be safer.

The funniest thing about mountain lions, to me anyway, is that they have domestic cat tendencies. Maybe it is just easier to see that some tendencies between the two are alike because I have a few cats. The point is that cougars are as picky as a house cat about where they walk. They do not like graveled roads with larger sized gravel. If they can help it, they will take a different path. Decomposed granite doesn't bother them but, pea gravel does. They really like to walk on the softest, most manicured of soils that they can find if grass or a road is not available. That just happens to be access roads, trails, paths, wet or dry sand by creeks, maintained rows through the oranges, grapevines, olives, and nut trees, and soft soil along the side of the road, minus walking through puncture vines. They do not like to go through soil that holds a lot of water, which can be quite deep around here at times. By watching how my domestic cat chooses the route that he walks, I see quite the same characteristics of a mountain lion. During the spring, summer, and early fall, I can usually pick up the tracks of at least two of our resident lions within about 20 minutes without the use of radio telemetry.

September 2nd, 2011.   My research is so interesting to me that sometimes I forget that there are people that don't like mountain lions. I find their habits interesting because they are so basic but every once in a while I get a surprise. It's always nice to see that Sneaker has come back around because his life expectancy as a male is very limited. Males tend to get into more trouble or seen before a female is seen. That's another thing. When I found his sent marking and tracks, I noticed it was harder to find tracks of the lionesses. They decided to hide. Since that day, I haven't found anything that would identify Sneaker in the area. I wonder where he went.

Chapter XV
Minus One Cub

September 4th, 2011. Something is not quite right. One of Hookah's cubs is missing. There are usually three sets of tracks with Hookah, and now there are only two. The little cub that is missing was the straggler, Jeremy. The life span for a cub with a mother is greater than that of a cub without. It is possible that a coyote may have killed the cub. Sneaker could have found Hookah and her cubs, but if that were so, then he would have killed all of them. However, he is the dominant male and had mated with Hookah, they were his cubs. I think she was in a different location hiding her cubs. It is possible that Jeremy got sick, and died. The last possibility is that a new human resident that lives off the land may have seen him and, shot him. For that matter, it could've been anyone in the area. Though these are just possibilities, one of them did occur. I just don't know which one it was. It can be difficult for a lioness to hunt enough prey to support three cubs but, I just don't see this happening with so much available prey in the area.

Finding the tracks of Hookah and her small family is often amusing to me. There was always one cub that would lag behind just a little slower than the others. The cub would end up trying to hurry to catch up. I saw this when I looked at the stride of the tracks. At first they meandered in a pattern where each foot had its own track. Then I would see the pattern grow into a straighter line and the hind foot would overlap the front foot. The next thing I'd see is that all of the cubs' tracks were together. The tracks are small yet I believe that the cub that didn't make it was a male. The tracks that are left are those of most likely a male, and a female. To tell one cub from another between the litters of two lionesses, I named Hookah's cubs Clause, and Celia. A male's toe track is different than females. As they grow I will see for sure which sex they really are.

September 20th, 2011. Zone 2 has a smaller riparian area that is inadvertently part of a larger one. This area is most frequented by both lionesses and their cubs. While visiting zone 2 at the back of the lake I found evidence by track of Sneaker along with Hookah, and her two cubs. I wasn't sure if it was Sneaker at first but then I looked closer at the tracks, and then confirmed it was

him. During a check of the smaller habitat site I found tracks of
Suka and Henry. It was cute, I could see mom walking along on
her way to some destination with her cub following along behind
her. These tracks were fresh, only a few hours old and they were
made in loose, dry, undisturbed soil. The worked soil or soil that is
driven on is so powdery in this area that during the hottest part of
the summer a swift breeze could have distorted them. In any case,
they were undisturbed.

The tracks were heading out to the east. It was clear that they
were leaving their favorite habitat. Perhaps they were taking cover
in the olive grove at the end of the avenue, until it was dark. They
would have to cross from the olive grove into a pasture for cattle
grazing with short grasses. That would make them vulnerable by
being clearly visible to the human eye. Waiting would have been a
smart play on their part due to the fact that the rancher that lives on
property across the road would have been able to see them easily at
4:00 in the afternoon. The olive groves were planted many years
ago. The trees are over grown with 90 to 95% canopy cover. The
ground is cool, and the fallen leaves are soft to walk on. Irrigation
on this plot is still done in furrows which also provide ample fresh,
running water for drinking, and the cooling of their feet. The
olives here are not managed the way farmers manage this type of
crop today. Currently, olives are grown as hedge crops because it
is more cost effective to do the harvesting.

September 23rd, 2011. This afternoon around 6:30 p.m., one of
my daughters and I drove out to a spot that I frequently find tracks
of Suka and Henry. On this day we found more of their tracks. I
left one camera. Going home, we stopped at the cat door along a
fenced part of the creek bottom. This is an opening in a fence line
that bobcats and the two lionesses use to either enter or leave a
brushy corridor. There were more tracks of Hookah and her two
cubs. From what I could surmise, it looked like mom had been
quite busy in pursuit of a small pig. The ground was hard. The
prey was in a hurry. The scuff marks on the ground told me that
she and her cubs had dinner. The tracks were most likely from the
early morning hours or the night before. I had found where the
cubs were playing, scratching the dirt like they had been attacking
a leaf. Note: Suka's cub is now approximately nine months old.
When I guesstimate Henry's age, I add on two weeks to be safe

because I don't have an exact week that he was born. It is now close to the end of the month.

Hookah's cubs are now about the age of four and a half months old. It is possible that she either hides them away in a safe place while she hunts, then after she's killed her prey; either calls them or goes and gets them. She may also have had them traveling with her waiting in a safe place until she gives the okay for them to come out. By the tracks it looks like perhaps they just move from one spot to another and on the way they find a meal. I do know that she was hiding them in the tall grasses by the lake for cover when they were smaller. The lake is essential to the wildlife in this area, and not for public use.

Getting pictures of a mountain lion is really hard. I have had more bad pictures than good ones. There was one picture, even though it was not clear, taken by the lake that I could see a mountain lion lying in the tall grass with two nursing cubs and one climbing around her neck. The shadows under the trees made it hard to positively identify them even when the pictures color was changed to black and white. There were certain things about that picture that were clear enough to distinguish parts of the figures. Sometimes their camouflage gets in my way.

Between the sloughs of the canal, I could smell a dead animal. The only problem is that with my back injury, I am not able to get down the ravine to see what it is. There were tracks of two cubs at the top of the other side of the ravine. I wasn't sure what happened here. I was worried that maybe something happened to Hookah or perhaps the cubs just weren't allowed to go down the ravine while mom had a bite to eat. I'm just hoping it wasn't Hookah that someone was eating while the cubs were on the bank. I wasn't able to pick up any cub tracks, but that doesn't mean they weren't down there. After examining the soil determining that there was a great deal of soil compaction, I didn't worry anymore.

A photo of the overgrown olive tree corridor in zone 2. I do love this place. The tree canopy shades them from the sun, and the ground is cool from irrigation water in the summer. There are plenty of places a mountain lion can disappear into the brush that lines Wahtoke Creek.

The runoff from the Friant Kern Canal creates ponds for bullfrogs, and crawdads. This is the perfect place for cubs to practice their hunting skills. The grassy, tree lined banks of the water's edge are cool with a slight breeze in summer months. Snowy egrets, gray heron, raccoons, opossum, beaver, gray squirrels, bobcats, and coyotes all frequent this area.

Wahtoke Lake in early spring of 2012. Campbell Mountain and part of Jesse-Morrow Mountain behind the lake.

Chapter XVI
Missing Lions

September 25th, 2011. Today was particularly disturbing. I had a hard time finding the tracks of any lionesses or their cubs. This was strange to me since they've all been so easy to find this season. There were no signs of Hookah or her cubs, and it was like Suka had vanished as well. When the habitat checks for the girls were over, I had a kind of empty feeling. It was like going home to an empty house, knowing that was completely out of the ordinary. This visit had left me uneasy. Maybe they were resting in the lake bottom. It's too early for their seasonal range change. I know that Suka is the youngest mature female in the area, so she would be the first to go back to where the center of her normal range is located. The problem I'm having with this is that she is leaving me nothing to go on. The avenues she travels have no tracks. There can be only one reason for this. Either Barbara Alice or Sneaker has come back but, I question this. Their times to come back don't come until the end of the year. No one has seen the gray lion since last fall.

September 28th, 2011. On agricultural land there are a lot of places I can drive to, and then set out on foot. I decided to do some back tracking today to see if I could explain Hookah's lack of presence. During mid-morning I went across the road to the empty pond area. I was relieved to find a poof of bird feathers, and it was about twenty feet away from one of Hookah's old kill spots. Hookah had been here. I also found some pretty fresh lion scat. Hookah had plenty of food. I found tracks from a sow with piglets. This seems to be her favorite meal next to birds. The kill spot though was her signature. I was happy to see that she'd been here but, I think she and her cubs were only passing through. From what I see from Hookah, I know that she has a specific area that she travels most. I also know that when sharing a range with Suka, their ranges overlap. Hookah and Suka do quite well working out their overlapping space on a daily basis. However, there are parts of their territories that don't overlap. These parts are the major areas of their sections. It's hard to describe what their areas look like on paper without providing a map. Through tracking the girls I've seen how mountain lions do not walk in a straight line, they meander from one point to another, back and forth to each place they visit. The points are spread out with considerable distance. Each lion travels their own trail intermittently crossing into the others habitat, and back again. There are fixed points for each of the lionesses, and for Sneaker. These are places where only their tracks are found. Each habitat has one or two points like this. The male crosses over into each female's habitat but, the females try not to invade each other's space. There is only one section I've found where they actually walk over the same bridge.

The communication between lions can be seen in one particular fixed point. This was a wow moment for me. I had never seen this before. I was used to finding each lions scat at specific locations, and without another scat in such close proximity, unless it was from a cub or two. There is a section of land that offers, cover, food, water, and a form of communication. There are three mountain lions in this description. Two of them are female, and the other is male. These three are mature mountain lions. There are two cubs belonging to one lioness, and one cub belonging to another. The area of description is shaped as a triangle. One section of the canal has three bridges. Sneaker would leave scat at the entrance of one bridge closest to the mountain. Down the canal bank road, at the second bridge, there was a scat from each lion or lioness that had passed through over a very long period of time. There were absolutely no bobcat scats in the mix. At the end of the second bridge right before the pavement were always tracks of each lioness, and her cubs. Along the road were the scats of all lions past, and present. At the access road which crosses between a thistle field and an irrigation pond, is the scat from Hookah. She is obsessive compulsive on how she leaves her scat. This scat is also directly in line, though no straight line was followed, with her scat on the edge of the third bridge. The access road between the thistles and the pond, half way down, was an area for another cat scat.

There are three lines of travel involved with each mature lions scat. By track I could see that each cub of Hookah, the first time she had taken them there through the third, would meander around like cats do, and wait until mom chose to do her business. Then they would find a spot away from hers to do the same.

Interesting fact: This is another occasion where mom has taught children the rout to travel, and claim. Further note on Suka and Henry will be given later. Sneaker can find any lioness from this communication even when the girls have cubs. The lionesses hide their cubs from a male yet, when the male is not in the area they will visit this spot on their way through. I think he knows they're his and won't bother them.

Something else about this is that I think Sneaker can tell by scent how old the scat is giving him an estimation of time when the lionesses and their cubs were last there. Like domestic cats, mountain lions can see UV to tell how old the scat is. This also lets him know how many cubs are in the area, and maybe who they belong to. I think the girls check these locations to make sure he's not around before they leave their scat or let their cubs leave theirs. I also think that it works for the girls as it works for Sneaker. They can tell by scent or UV when the last time was that he was in the area. This gives them time estimation on when he'll be back. This may indicate an early seasonal range change when they have cubs. Then they start burying their scat.

There is one other point like this one that I've not cataloged for the purpose of their safety. It lies close to a residential area. The scat lines the road in a section about three yards long. The scat communication is between a male and female. This marker had lasted for a few years until someone decided to tractor. Now it's not there. This could be further marker of willingness to mate between two mountain lions. The scat was not fresh by both parties every year. I have more information but, that might put the cats in danger. There is one lioness that doesn't spend the majority of her time here but, comes here for breeding because this is where she knows she'll find Sneaker. Is this when Hookah and the other's change their ranges earlier, than when Barbara doesn't come home? When I say earlier, it's only by a couple of months. This may be my answer. I think Hookah and Suka were dodging the presence of Barbara Alice and, since Sneaker was back they had to keep their cubs out of sight. What better way to do this than to leave the territories where their boundaries would overlap with Barbara's? Hookah was in a tight spot. Her territory had to change to encompass an area off her normal, seasonal territory.

September 29th, 2011. While trying to piece together the disappearance of the girls, I look back a few months earlier when Barbara Alice was seen crossing Central Ave. heading towards the Kings River in May. I mention this because of the spray noted on a bush. The spray was from Sneaker with matching tracks. I have learned something new about Sneaker. He comes back in September as well as in January! Sent marking for mountain lions is a calling card for finding one another. The cat screams heard near the area that was scent marked could mean that they have found each other since the screams were short lived. Her return could also be because the month got cooler. It may have been time for her seasonal change in range, decrease in prey or mating cycle. Thought- If Barbara Alice has had cubs over the summer, she would be hiding too. There would be tracks from her, and her offspring in a different location. There is nothing to the effect.

September 30th, 2011. The night was silent. All of our animals were sleeping. It was about 11:30 p.m. when out of nowhere I heard a cat scream. It was definitely not a house cat, but a very big cat. That is a sound that makes my skin do funny things. The scream came from across the road in the bottom pasture. The sound of a mountain lion scream can travel through the creek bottom for a long distance. There's no way to tell just how far away she actually was. No dogs gave off a warning sound which would help me establish direction and distance.

October 1st, 2011. Around 4:00 a.m. I got up to make coffee. My husband, the skeptic, and I say that lovingly, meandered out of our bedroom, and looked at me. All I could muster to say at this point was, "What?" Tracy asked, "Did you hear that?" I said, "Hear what?" He asked, "You didn't hear that cat scream?" I said, "No, I must have missed that." He couldn't believe I didn't hear it because he said it woke him up. OK, so I needed coffee before anything else.

October 3rd, 2011. Thought- From the 16th of August, to the 2nd of September Suka was alive for sure and so was her cub, Henry. Their tracks for the most part were always together. On September 22nd, Suka and Henry's tracks were fresh. On October 1st, Suka's tracks were found but I didn't find many or at least enough to feel comfortable. Most of the ground is covered with leaf debris from falling foliage and tracks can be hard to find. Usually, the two travel out of an area together but there were no exit tracks visible on this day. Henry can survive by himself now, but he still needs his mother to help him perfect his hunting strategies through play. This is one of the things I've learned about what a lioness does with her cubs. I learned it from Hookah.

In regards to the trip up the mountain that Tracy and I took in November of 2010, I thought about the two different encounters with Hookah and Ben. Hookah had let Ben roam the top of the mountain, without her being close. Ben was fifteen months old. Hookah was quite a distance away from Ben, and I'm sure she didn't know about the encounter he was having with the guests we'd brought up there. The only differences between Ben and Henry were their ages, and the locations for which they'd been raised. The locations are important because Ben was secluded from people. The only danger he would have encountered was that of a coyote or a male lion. With Henry, the dangers exist of some really big coyotes, the occasional field worker, and the riders that pass through zone 2.

In my mind, I'm still asking, "Why," about the separation between Henry and his mother. I find his tracks more often than Suka's now. The girls have shown me that it is not the norm to be too far away from their cubs at any time. I looked in a 40x40 radius around Henry's tracks and, still found no sign of Suka. Then I looked in a much wider span. There were still no signs of Suka's tracks at all. Perhaps Suka is letting him hunt independently yet, still under her watchful eye. Henry and Suka's tracks are not in zone 2 as much as they were. The possibility of Suka taking Henry to the citrus groves that surround zone 2 and 3 is probable. The south east end of the study is the edge of zone 3, near a chain of foothills that border tree fruit.

Chapter XVII
An Unknown Presence

On Wednesday a huge storm is moving in. The soil type in all four zones will become muddy, slippery, and in some areas, deep. My daughter Kate has, so kindly, volunteered to help me check for a carcass near the canal outlet. She will be going down the steep ravine between the two canals by rope to see what kind of carcass is visible, and producing that rotten smell.

October 5th, 2011. When Kate came home she was ready to go look for the carcass, so out the door we went. We were so excited we forgot the rope! It was okay though because we stopped by my parents' house and were able to get one from my father, who thought we were nuts for going to look at a carcass in the first place. I asked him if he wanted to go and with a smile on his face, declined the offer. Bryan was as enthused about going as my dad was. They stayed at the shop smiling, and waving as we left. When we got to the spot where she needed to get down into the ravine, I warned her about the rattlesnakes. Kate looked at me. I knew she wondered what she had gotten into. I gave her a really long stick and told her to check the brush before she stepped into it. The look on her face was priceless. There were no snakes but, it's always best to keep them in mind. If I really thought there was danger, I would have written the excursion off and hoped for the best. She makes me laugh.

When Kate was down at the bottom of the ravine, I really wished it was me that had gone down there. I wanted to see everything that she was looking at. I asked her a bunch of questions, and each time the answer came back that there was no sign of any carcass. She searched around everywhere and found nothing, not even teeth or small left over bones. As Katie stood down at the bottom of the ravine looking up at me, asking what's next, I had no answers. All I could do was ask if she'd looked under those bushes. Her reply was, "Yes, I've looked everywhere and there's nothing." I asked her if there were any tiny bones or teeth. Kate got a little snippy and said, "I know what teeth look like Mom!" I asked her to walk back towards the beaver dam. She still couldn't find anything. It was clear that the trip was over. Kate came back up the steep hillside of the ravine pretty proud that she'd braved the rattlesnakes. I was really glad she didn't find one. The fact that she went down there in the first place was touching to me. How many kids would help their mom look for a carcass, and face fear at the same time? Just mine! Kate has been my right hand for years. On the other hand…How could it be gone? Where did it go? Well, something with four legs must have dragged it off. I wanted to make sure that it wasn't Suka. No such luck. I'm not sure what it was. I really don't want to think that someone may have tracked her down and shot her. After investigating further by talking with people in zone 2, I realized that there are five different people that ride through this area on horseback. Only three of which are ranchers.

Later in the afternoon, I went to the pond in zone 1, behind our property. The olive trees provide cover all the way around the pond. The cat tails, reeds, and water grass provide the rest. The cat tails and reeds are home to bobcats, ducks, opossums, and raccoon. Coyotes use it as a safe haven during the evening, and early morning hours. This area is used by three different lions for water, food, and cover from early spring to late fall. Several lion scats among coyote and some bobcat scats, have been noted. The latest scat for this area is from a lion. It was a single lion just passing through, there were no kill spots found. The soil was so compacted that I saw no viable tracks. The scat mostly marks the north end of the pond. The newest lion scat was in a minor state of decay with the hair and bones of a rodent or bird.

The same day- Katie had gone with me out to the slough I just mentioned. She was following me in the truck while I tried to find clues of who this lion was, and where it had gone. She was sitting in the driver's side of my dually when out of nowhere came the sound of a big animal moving fast, and in my direction. I stopped to try to figure out what it was, and where it was coming from. I knew it had four legs but, that's about it. Into the truck I went, just in the nick of time! A big gray Doberman went flying past me at full speed! As it passed the truck we both stared at it. Katie turned and looked at me; she cracked up laughing at me for jumping into the truck so fast. I couldn't get all of my body into the truck because it still hurt to bend. I figured if it were a hungry animal, it was only going to get one leg. She said, "Wow MOM, it's only a dog!" I laughed at myself but, I didn't know what the heck it was, and I wasn't taking any chances. Right after, a man on a horse came by. He asked if we'd seen his dog. He said, "Which way did it go?" He probably thought we were nuts because we laughed at him when he asked us. We just told him, "It went that way!" This was the last event of the day.

Thought- The girls are hiding with their cubs. Within a six and a half week period, I found lion scent marking, and cat screams were noted in two different study locations. In applying what I have learned about them so far, the mating cycle of a lioness from time of mating to the birth of cubs is about ninety one days. Sneaker would have had time to mate with one or two females. It is possible that if there were only one female that it could have taken a week or so to find her. The only problem with this thought is that I can't find any tracks. Perhaps, I missed the tracks of the first lioness due to the fact that without permission from the land owners to be on the property, I couldn't go there. I was only able to survey the road side from one location leaving three parts of the perimeter of vineyards, and citrus out.

The farm land to the north of Hookah's winter home was not tracked at all. I avoided asking permission because I didn't want to set off any alarms in the neighborhood. The road along the opposite side of the lake is accessible to me. However, the downfall of having to participate in human interaction regarding a mountain lion may in turn impede the lions travel or life span. Females are said to breed at approximately two and a half years, and males do not scream for a female when it's time to mate; only the females scream. I have not seen anything in text stating that the maturity of a lioness is definitely two and a half years but merely an approximation. I think I have reached a conclusion. There is only one male seeking to mate with one female at this time, and they traveled within two separate study zones. With this information in mind, we may have new cubs in about three months, around the first half of February. The lioness could be Yinka.

Note: 1/27/12- To date; there were no new lioness tracks anywhere. Given the distance between mating calls over a period of a couple of weeks, a lioness will travel to find a male. The lioness must have mated with Sneaker, and then traveled a distance away from him at the end of her breeding cycle. This gives him the message that she was ready to be solitary again. She wouldn't go too far away from where they'd mated. She'd then stay in that area, clear it of predator's, then ready herself for

101

denning time. Due to overlapping territories, her home range would be to the north west of the study. There's a large mountain chain that separates that location from this one. If there was a possibility of this female being Hookah's first born, if she is two years or so older than I think she is, it could be her first cub was a female. This would mean that Hookah has had more than one littler in this area, and they all come back. There is one other thing to keep in mind. From what I have heard (not seen), when Sneaker is chasing deer or pigs down from the foothills, and he has a mate with him, even though they are different mates in different years, it could have been Barbara Alice after all. She was seen within the same time period as Sneaker, when he was allegedly shot.

The information I have regarding sex of mountain lions is three adult females, possibly one mature unknown transient female, two subadult males, one subadult female, and one primary male sharing their overlapping home ranges. I haven't seen Barbara Alice here but, the mating calls and reported sighting of her may mean she is here. If she's not, then this gives way to an unidentified presence. Upon the time of disbursement, four subadults will leave. The question remains: Who is the lioness? Is it Barbara Alice or is it Yinka?

October 14th, 2011. The afternoon was cold so, I got in the truck and turned on the heater. I went out to retrieve my IR camera. It was gone! I know several people had seen the camera. Even though there is always the possibility of this happening, it never has before. It is disappointing to know that someone, probably someone I don't know, would do this. However, it is always a possibility so I just suck it up, buy another one, and keep going. The results are always worth it. I just hope that none of the lionesses or cubs pictures were on the SD card. I would hate to think that if someone got a hold of their pictures, that it would endanger them. I will eventually find out who it was.

October 15th, 2011. Fall is here. The weather has turned cold. Right now, cold is not my friend. On the twelfth day of no tracking, it decided to rain. As I sit inside looking out the window, I wonder if Hookah and her cubs have really left or if they are holding up somewhere out of the rain on the valley floor.

Yesterday, I had found some faint tracks of a lioness, and the tiny tracks of one cub. Did I miss a set of tracks, was this Hookah or was it Suka, and Henry? I know where these tracks lead to.

They go off into an orange grove, back to a part of the empty canal that still has some water ponds with remaining small fish, crawdads, and frogs. The gray fox like to hunt here, too. It is possible that Suka and Henry did venture over into zone 1A, by way of the lake. It would be nice if they did, at least it's safer for them here. Suka's might be trying to avoid any contact with Sneaker because of her cub. This orange grove is fenced and provides a safe haven from people during this time of year.

I went to the pecan grove and talked to the head man that sees the cats early in the morning. I asked him if he had seen Suka and Henry, together. In an enthusiastic tone, he told me he had. I was relieved. The man did confirm my thoughts; he said he hadn't been seeing them as often. I wonder more about Suka because she's younger than the others. Each mother lioness governs her cub(s) closely but in doing so; they do have different habits with what they allow their cubs to do. Hookah keeps her cubs with her or she remains very close to them until they're old enough that she can give them some room. This only happens when they are on a mountain top. Whereas, Suka is a first time mother, and I would think that she would not be so relaxed with Henry's freedom. Her choice to let her cub roam further away from her may be due to inexperience. It could also be due to a naughty cub that likes to wander when he's not supposed to.

There is another part of the canal that handles the runoff. It is shrouded in trees, wild grapevines, and brush. The adjoining farm land is primarily almonds, and stone fruit. This place is well protected and there are several escape routes the lioness and cub(s) can take if farming equipment should enter the area. Culvert pipes are an option for them to take cover in if necessary. The canals are now empty. In some places along the banks, the side slopes are gentle. The canals provide routes to travel in this area. One specific canal is a travel route to the cattle grazing land across the road, through the citrus groves, and down the embankment into the old creek bottom.

When the canal is empty, the water gate is left open. A lioness and her cub(s) can freely walk from the canal into the slough without being seen.

The tall grasses and overhanging tree branches will give seclusion to the lioness and her cubs. There are plenty of ducks, crawfish, and rodents along the way. Fall is my favorite time of year. The leaves on the trees change color, the days get shorter and nightfall comes earlier. The coyotes take full advantage of the early evenings, congregating for their hunts. When Barbara Alice comes back for her visits, the coyotes know she is around so they don't stay to hunt in one area for long. However, it's like that with all of the mountain lions. During the last week that I heard the lionesses mating calls, I monitored the songs of the coyote. I could tell the direction they were headed after the third yipping phase. The yipping intervals grew wider apart until they were so far away that their songs became faint off in the distance. Some nights they do not call at all. When the wind machines turn on, they just hunt silently in darkness. I think this may be because the wind machines do not inhibit the travel of mountain lions; Sneaker has become accustomed to them. When my daughter saw the faint figure of a lion in the past, the wind machines were running.

Funny, but I think that Sneaker makes stays longer, down here in the valley because of the fog. The fog has a certain charm if you're not driving in it. One of my most favorite memories was to watch equine yearlings running in a pasture when the fog was so thick that I could hardly see five feet in front of myself. Unless people live in the San Joaquin Valley, it is hard to understand what I've just said. The Weather can aid the disguise of a mountain lion.

Some years ago, I worked on a thoroughbred ranch. We normally began work when it was still dark. Flashlights lit the way but, on this morning the fog was thick, and the darkness didn't help. I went out with the crew to a distant pasture. We had to find the yearling's for their morning check-up. It was important to keep sight of the person in front of us or risk losing them into the fog. A person had to know the way quite well because the fog can disorient a person, and make familiar locations deceiving. Once we reached the pasture we climbed through the fence, staying together at least an arm's length apart. The sun beginning to come up brightened the fog just enough to where we didn't need to use flashlights. As we walked through the pasture, we clicked, and called for the yearlings. There was nothing, no sound at all, just the dullness that falls over a foggy landscape. Then out of nowhere I could see the heads of several horses silently appearing with dark, wet manes flying in the mist. The movement of their bodies was fluent yet, there was no immediate sound. In just seconds the sound of thundering hooves passed by, then their sound was gone as quickly as it came. The horse's movement did not correlate with the sound at first. Several passed in front and in back of me, some dodging me. The movement of the herd cleared a hole in the blanket of fog. Other horses off in the distance were partially visible. The stampeding yearlings were only feet away from us. In what seemed a split second, the herd of horses moved away, their hoof beats silenced, still with manes flying, disappearing into the thick fog again. If the sound of a herd of thirty horses at full gallop can be silenced by the fog, and their appearance hidden from view until they are upon us, imagine what it does for mountain lions anonymity.

During October and November, the mornings are cold, and wet from dew. Most of the time a mountain lion will move to higher elevations. There is more chance of finding prey when the weather conditions are warmer. Ground nester's like the Road Runner, Meadow Lark, and other prey species' like rodents, go about their daily business when the sun comes up. This makes finding food for a lion easier.

Suka seems to prefer the neighbors Rhode Island Red chickens, bullfrogs, and pigs. Unlike Hookah's hunting signs, I do not find poofs of feathers when I'm on the trail of Suka. Recently, I found her tracks; she had been in pursuit of a deer. I was quite impressed when I looked at the tracks. No matter where the deer tracks turned she was two steps ahead. When a deer runs, it pushes with its hind quarters for a longer stride and loses time getting away from a lion. A lion pushes with its hind quarters but gains more speed because of its musculature, length of body, a tail used for steering and balance. The tracks of the final skirmish were definite, she got her dinner. The tracks of Henry were present, and I doubt he was far behind on the hunt.

From all that I've seen of Suka's tracks, I've not seen her on the trail of a deer before. Mule deer congregate under the fig trees down in the pasture. If she's been hunting on grassy ground, it's highly possible that I've missed them. I know that a mountain lion can pull down an animal much larger than itself. I had hesitated to give this much thought because she was so slight in weight when I'd first seen her. I had overlooked the fact that she had grown. Thinking that her size and weight would inhibit her will to hunt prey larger than she was at the time, was an oversight. Then realizing that months had passed, about six actually, this definitely would have been enough time for her to put on more bulk since she isn't nursing a cub anymore. With this in mind, I realize that aside from Suka's small build, a mule deer would've been the perfect size for her to hunt. In comparison between the two lionesses, they were both on the slight side because they had been supporting their cubs. When I look back, I see the difference between their weights from then until now; they've had a huge growth spurt regarding bulk. Hookah is one big kitty. She was a happy lion, and seemed quite content as she sat atop a large boulder to watch us. When she sprawls out on the pond bank there's no missing her.

October 22nd, 2011. The howling coyotes at night are beginning to come closer to the house. The howling lasts a lot longer. During the summer with the two lionesses and their cubs, down on the valley floor, the coyotes were pretty quiet. I would hear one or two but, now the whole family is out and they're looking to join up with their relatives. I hope that Suka and Henry have left for their range change to the foothills. I haven't seen them. Hookah and her two cubs are the only remaining family in the area now.

After coffee, I went out to the Pecan grove. A worker did his best to show me where he had seen Hookah and the cubs. He told me that she still has two cubs and not one, like I'd thought. The man has been an eye witness of these lionesses for a while. He arrives at work when it is still dark. He is jovial when he shows me where he sees them in the mornings, makes it plain that he keeps his eyes open, and tries to stay out of their way. He pointed me towards where he had seen Hookah with her cubs early the morning prior to our visit. I went over to the location and sure enough, he was right because there were moms' tracks along with her two babies. They had been heading towards the slough where they could hunt ducks, and bullfrogs.

The reason I thought Hookah was missing another cub was based on the one set of cub's tracks with its mother that I saw in section 1A. I mistook these tracks for Hookah's because sometimes she visits this area when trying to avoid agricultural work. The size of Suka's' footprint is similar now to the size of Hookah's, and especially in the dryer soil. I should have measured their tracks every month using the minimum outline method to keep a record of their growth. Instead, I only used it for identification in the beginning. I didn't take into account that I might lose track of one of them. It also might have helped me age the unknown lioness. Next time I will measure more frequently to record this.

I had gone out to find another location to post a camera that would be directly in the line of where she normally walks with her cubs. The problem was that no matter where I put it in a pecan grove, it would be visible to the people that rode through on horseback. I'd never had trouble with any cameras missing before but, that one time was enough to make me try harder to conceal them better. The camera was placed in the open. I got some pictures; however, they were only of a couple of wild pigs. The time for posting cameras in the pecan grove had come to an end for the season. They needed to harvest. That meant that it would be even harder to get a picture of a lion because of the harvesting noise. Hookah will find an alternate route to travel, and it's getting closer to the time for her to take her cubs up to the foothills.

Booboo the bobcat is hiding in the orange grove taking a bath. I chased him with my camera for a while before he became bored with me.

Chapter XVIII
Life without a Mother

October 26th, 2011. The workers at the pecan grove stopped me as soon as I drove in this morning. They told me that on their way to work they'd seen a dead mountain lion on the side of the road. They told me where to look so; I went to check it out. I had to see this for myself because they told me this once before, and when I went to see if it was a lion it turned out to be a bobcat. The only reason I really wanted to see if this was a lion was because I haven't seen Suka in a while, and her tracks are not with Henry's lately. The men also told me that they have been seeing one cub walking the canal bank alone in the mornings. Suka used to walk with him every morning. This is a problem. I thought this may be the case. When I went to check it out, I found no sign of a dead lion or a bobcat near the specified roads. Just in case, I checked the other roads in case someone had picked up her carcass. The two roads that the workers directed me to encompassed the corner of zone 3.

Earlier, I had stated that the avenues she frequently travels had no tracks. I know why now. She would come down off of a low elevation mountain, and instead of traveling across the bridge, would slink down into the drainage ditch that meandered along the Friant Kern Canal. There is a small boggy section of acreage that has become overgrown with thistle, and hides the wildlife as they travel in daylight hours. The exit of this path comes out at a holding pond near citrus. She would then travel through the trees to cross the intersection that led to the adjoining foothill chain. Normally I do suspect people when I can't find a lion or lioness. If she'd been moved by a person, it's probably because no one wants a dead mountain lion on their property unless it's buried or out of sight. A common dump site for dead animals is the Kings River.

Still, no sign of Suka being dead is a good sign. I found nothing. I spent a good two hours looking for her. I am still hoping I see her again in the spring. I think I'm hoping for the impossible because had the dead felid been a bobcat, it would still have been on the side of the road. Life for Henry will be difficult. He's not supposed to have to grow up without a mother. The odds of him surviving are pretty slim but, Henry is not a tiny cub completely dependent upon his mother. Though his hunting skills are not what they should be right now, as he grows he'll slowly gain more. The fact that he has some experience hunting small game may save him. He may survive on his own. Henry is nine months old now, and he'll have to live by his wits. Suka will not be there to watch over him while he sleeps, snuggle him to show him he's loved, wash his face or purr to him before he goes to sleep. She will not be there to help him catch his meals if he fails. Henry will have to learn to keep himself safe, and hunt well in order to survive. Coyotes will be able to make a game of him now whereas before, he had the protection of his mother. A young mountain lion cub's life without a mother is in no way a fairy tale. There is no father to step in to help take care of him. On the contrary, he will have to be on his guard in case another male lion smells his presence anywhere. Henry would surely loose this battle. Life will be harder for Henry. It is a sad fact that he may not make it alone.

I am not a person that shows a great deal of emotion where sadness is concerned but, there are moments when I'm looking for Suka and ultimately know that she's not there… no matter where I look. Once in a while when I'm in those quiet places that she once roamed, I listen to the same sounds she did, and I miss her. I feel my stomach sicken sometimes. Yes, even for a wild animal that could make a meal out of me, I feel sadness for her untimely death. There is no more hide and seek to play with Suka. I'm sure Henry misses her, too. With each lion that goes missing, I feel that there is less hope for humanity.

October 27th, 2011. I went back to the olive grove where I saw Henry's tracks. I found them but, they were not alone. A really big coyote had been tracking Henry. Keeping in mind that all wildlife follows another, I tried to keep it in perspective. Maybe the coyote would lose interest and find a rabbit to chase. On my visits to this zone in the days to follow, I found that this was not the case. This one coyote was determined to find Henry. I'm really glad he can climb trees. I think that the coyote will have bitten off more than he can chew one day, because Henry will find him instead.

October 28th, 2011. Today it was called to my attention that when I go out to look for track and sign of mountain lions to remember that there are areas of land that may be perfect habitat for lions but, also perfect habitat for pot growers. There are a few areas that I enjoy tracking because even though there's easy access, they are secluded. The area where I had a camera stolen was like this. I liked putting the cameras out there because the habitat was great for lions to hunt. They had everything they needed. Suka and Henry really liked it out there.

The day I found that my camera had been stolen, I was standing not far from where I had it posted. I could hear a battery operated radio out in the bush somewhere. When I left, I drove back into the trees a little farther to see if there was a farm vehicle out there. There wasn't. Everywhere I had been in that particular section, I had not seen any drip system. There was a section of brush that was cut for a human to fit through, down into the brush. I went home to use the computer. I could see what the terrain looked like from an aerial view. There were spots that were open, and sunny with water near. Other areas were so covered by tree canopy that I couldn't see the ground. Well, we will have to see what's out there. Perhaps there is more to that area than meets the eye. There could be more places that Suka and Henry would use rather than just what I can see. Keeping safety in mind, I try not to investigate things by myself in case I can't get back to my truck.

November 4th, 2011. Today, I went out to zone 3 to set cameras as an attempt to see if Suka may have taken Henry back to the mountain. There was a lot of bobcat scat around the base of the rocks. I walked to the location where I had seen Suka and Henry for the first time, to see if they might be around. No such luck, there was no lion scat or tracks anywhere. It is so much easier to see them than to get pictures of them. I have to shake my head sometimes because when I don't have a camera near me, is the most likely time I will see one.

November 6th, 2011. Today was a really pretty day. My daughter, Melody, and I took her dog, and went out to the slough to see what we could find. We took her dog because he needed some one-on- one time with his mommy, and because he sees and hears things that we don't. We had no plans as to how far we would go. We parked in the pecan grove, and then walked along following the canal bank. Red was so happy, he couldn't believe he was getting to go explore with us. We walked down the bank of the canal onto an access road. Red disappeared through the wall of trees that surrounds the slough. We heard him bark, and go after something. Whatever it was most likely ended up in a tree. He came back to lead the way for us, of course he thought he knew where he was going. I couldn't get out to the part of the slough where he had gone. I don't think he treed Henry because he came back to us so quickly. I worried about this. Though, if he had treed a lion, I think he would have stayed and barked. We probably would have heard the cats' contempt for that situation if there'd been one.

The slough was to the west of us, and olives were to the east. The olive trees had been flood irrigated. The water ran over the road making a huge mud puddle. This is always nice to find because wildlife likes to walk through silty mud, leaving tracks. There were a lot of raccoon, and opossum tracks but, no cat tracks. The road lead to a large open, green grassy field, and stopped. Along the border of the field and the slough, a lion and a bobcat left their scat to mark their visits. One visits and marks, then goes away. The other follows, and does the same thing. No one ever wins this quiet discussion unless one eats the other. In finding this clue, I knew one of the cats had been there.

Mel and I decided to go into the brush as far as we could. There were a couple of fallen trees and some vines hanging down. The ground was covered with leaves and branches. While we made our way through the dense shrubbery, and deep debris of dead foliage, I ran across a fallen tree. The roots had pulled up when the tree fell, making the perfect den for some large animal. It was wonderful habitat for any kind of wildlife. I looked up into the trees with my binoculars to see an owl sitting close to the bark of a tree, sleeping. It was a screech owl; its face was partially blocked from my view. I tried to wake it up by hooting, but no such luck. The owl shunned me. The beaver wasn't around but, the dam was still intact.

There is a section of the slough that is so overgrown that it's hard to get into. The bases of trees are bent, and wound around, lying close to the ground. The thick limbs are low making it easy for any animal to walk up into the tree to sleep. Cob webs are a given. They dangle from overhead. It's necessary to watch our footing because falling or twisting an ankle in this vegetation is easy to do. When we got passed the overgrown perimeter of trees and vines, and into an open section where grass and dried weeds stand as tall as my waist, we could see so many colors of green surrounding us. If I didn't know which state we were in, I'd swear there was a gator down in that mossy green water.

There were no signs of any big cats just the tracks of one bobcat, some opossum, raccoon, and birds. Near the end of our visit I found a sign of life from Henry. He had left the fluffy left over tail of a squirrel. I have seen this before; a lion shucks the tail off of a squirrel but eats the rest of it. I know that a bobcat will do the same thing but, being that the scat and the tail were so fresh, I attribute both to Henry. This kill had to be Henry's snack. He marked his territory with a scat pile, too. When he would travel with his mother, he would leave a scat pile not far from hers. I've noticed this with Hookah and her cubs. They do the same thing. The moms are teaching their cubs to mark territory. There was something different about this scat sign. Henry knew his mother wasn't coming back. Instead of leaving his scat where he would normally leave it, he left it exactly where hers was. Henry has taken over her habitat. As far as I can tell, he hasn't ventured back to where she was killed.

I cannot imagine what he did at that moment or the distress he must have felt. To have to leave his dead mother must have been so scary for him. What a brave little lion he was to turn and leave her, then travel back to the last place that he felt the safety of having her.

When I go out to survey zone 2; it can be a little eerie. This is a place where wildlife takes advantage of hiding in shadows. When the wind rustles through the leaves on the trees, it draws my attention. I listen closer to see if I'm hearing an animal's movement instead of just the wind.

November 15th, 2011. Today my sister-in-law, Sophia, called. It is always good to hear from her, I enjoy her personality. She said she'd been thinking about me. I asked her, "Why is that?" She told me she had been at her kitchen window and saw something large jump down out of a tree into the grass below. It was about 7:30 a.m. and she needed to get her glasses because she was quite frankly, a little stunned to see what she thought she was seeing. The large cat had a long tail, and this tail was visible when the cat jumped back up into the tree. While he was up there, a golf cart drove by right underneath him, and kept going. The lion must have heard it coming. After the cart had gone, the lion jumped back down into the grass to resume his hunting. Again the lion was interrupted. He jumped back up into the tree. It must have been a false alarm since no golf cart came by. The cat again jumped back down out of the tree to resume hunting the small rodent in the grass. It wasn't too long before he jumped back up into the tree again. Sophia didn't see the cat anymore after that. The poor thing probably needed a nap! The gated community they live in is near Avila Beach. The hill sides are beautiful. There are many sightings of mountain lions there. They too live in the gated community.

No one has had any problems that I know of. There are joggers and dogs, golf carts and people walking the trails, plenty of cars going up and down hills, bicycles zooming here and there, and a surprising amount of people that power wash their homes almost on a weekly basis. They do this to keep the salt at a minimum. These trails and roads are occupied all of the time. The proximity of the houses more than overlooks the trails in some locations; they are right next to them. It's a wonder that no one has complained. I was so happy to hear that she had seen this coastal mountain lion. She also saw how hard it had to work to avoid the public eye. I warned her not to walk her dog during what may be the cats breakfast time. There is a large percentage of quail in the neighborhood where she lives, and deer on the hill sides. It probably doesn't help matters much when she feeds large groups of quail in the morning. The mountain lion doesn't have far to go for its groceries. If I were to take a stab at managing this activity, I would implore people to "Not" feed the quail. This might reduce the activity of the lions within the neighborhood.

December 9th, 2011. There has been no sign of any mountain lions in the lower elevations of the study for some time now. I will continue to monitor fixed points to see if I can find any of them. Most of the work will take place on the tops of the foothills. In time to come it is possible that I will be able to monitor the activities of Hookah's young cubs, hopefully Henry, too.

December 22, 2011. Today I was anxious to get home and see what kind of videos I got. I was happily surprised. The videos were of a huge *Lynx Rufus*, he was beautiful and in full winter coat. The camera never bothered him he acted like it wasn't even there. It wasn't mountain lion footage but it will do for now.

December 27th, 2011. A family of three raccoons has made quite a bit of sport out of trying to get one of the cameras off of the mount that's next to a rock. They climbed all over it with their tails getting in the way of the camera lens. I wondered why they were all hanging around with a skunk. Later, I went back to trade out the memory cards, and moved the position of the camera. When I looked at the videos a second time, I had to laugh because it appears as though I have found nature's bathroom.

There was a rock removed from the ground for a landscaper. The raccoons, skunk, and a bobcat all use that hole in the ground to do their business. Yes, I have footage of nature relieving its self, WOW! I never thought I'd be sitting at my computer watching this.

December 28th, 2011. My son went with me today to check the cameras. We found parts of a large snake skin. There have been times when I've been skipping over the tops of rocks to avoid stepping down into the tall grass to avoid the possibility of a snake bite. More than once, I have stood on the tops of rocks and smelled the musty smell of a rattler somewhere around me.

I found some hair on the fence line at the top of the mountain. The hair was attached to the third row of barbed wire. Judging from the type of hair that Hookah has, the hair on the fence was the same texture and color. I was excited. This could mean that Hookah, or maybe Sneaker, had ventured over the top of the mountain. The cameras images that are unclear could be any of them. It is possible that one of them is hunting around on that hill top at a time when human residents of that particular area are all in their homes, warm and snug at this time of year.
Hopefully soon, I will find more out about this cat, and who it is. The grass is so tall, and thick that there are no tracks.

I know for a fact that Henry is still alive. I found his tracks in a location that was familiar to both he and his mother. At first when I looked at the tracks I was not sure if they were his. They were larger than the baby tracks I'd taken casts of. Then I realized that his little foot pads had grown larger. They are proportionately right for a cub of his age. They also have a distinctive characteristic the others don't have. Sneaker may be his father but, Henry is coming of age and is unguarded by Suka. Personally, I'm afraid for Henry. Most young male lions are killed by older mature lions in competition for territory. This case would be no different. The last casts of his tracks I got were in February, 2011. It is now January 2012.

It is discouraging to collect a camera to find videos that are completely black from the night. The secret is to look at the times, and dates on the videos. This can be a key factor in establishing the hunting hours of a mountain lion. When reviewing these videos, I look for the faint glow of red eye shine. Mountain lions eyes will flash yellow with a greenish tint. Sometimes they glow red but, most of the time, the pictures show a red tint around the rims of their eyes. It seems the lion can't help looking in the direction of the camera. They usually look twice. It helps to see the movement of the animal to tell whether or not it's a coyote or a lion. Usually a coyote's eye shine is yellow with no red at all, and their movement is quick and choppy compared to a cat.

Two coyotes in the Wahtoke Lake bottom, on a sunny day.

Chapter XIX
Seasonal Range Change

January 7th, 2012. Ashley came home for a visit. I took her out with me to check cameras. She remembered what I had taught her about tracking and had gotten better with it since she's lived out of state. We hiked up to the top of a foothill in zone 3. Since I don't have access to the adjacent foothill at this time, all we could do was use binoculars to see if one of the lions was basking on the rocks. We saw small birds fly up into the air like something had disrupted them, but soon they landed in the same location. We didn't see anything out of the ordinary so we called it a day.

January 12th, 2012. Today, Ashley and I went to retrieve camera memory cards. There were over one thousand pictures, and videos to scan through. The only wildlife for the week was one bird. When I go out to the foothills, I walk on the tops of the rocks to get to the cameras. It's a good thing I do because while standing on a rock, I caught a whiff of a rattle snake. It must have been a big one or more than one because the scent was really strong. They generally have a strong, musty smell. I couldn't see it but, I definitely knew it was there. The grass grows so tall that the rocks are almost concealed making it possible for a snake or a den of snakes to be hidden from view.

Ashley was following behind me. I put my hand up in the air to make her aware that she needed to be still. With no further warnings of a snake, I moved forward. Ashley got to the same rock, and stopped. She said, "Mom, I can smell a snake." I told her I knew, that's why I'd stopped. I told her to wait there a minute to look around her feet. She did, and since she couldn't see any immediate danger, continued on to my location. I've always had a hard time keeping shoes on that child. I laughed at her and said, "I bet you're glad you wore some shoes today!" She laughed, and agreed that flip-flops wouldn't have been a good choice. She also said she noticed that rattle snakes smell different than copperheads. I thought it was pretty neat that she'd figured this out. They do.

January 17th, 2012. Kate came down from the mountains to visit us. We went for a drive to zone 3. On the backside of the lower elevation foothill, she and Ashley followed me in the truck while I looked for tracks. I found them. The tracks were Henry's.

I found them in the same overgrown field that his father, "Sneaker," used to hunt pigs in. Sneaker used to cross over the mountain top, go down through the grapevines following an access road. The road bordered a residence where horses would run in their pasture giving him away. He'd then cross over a small paved road and disappear into the tall weeds. I think he liked hunting coyotes back there because there were so many. Now Henry considers this a great place to hunt.

The girls decided to walk with me tracking one side of the decomposed granite avenue, while I tracked the other. While following Henry's tracks, we found the first scat from him that I'd seen in a while. He had marked a boundary at the cross roads of the abandoned acreage. This was great news. He remembered where his mother, Suka, had taken him as a cub and stayed within the habitat. His feet have really grown. They are not the little foot pads that they were when he was five months old. They are more than three times that size. I can imagine what his feet would look like if I were to see him. Perhaps soon, I'll get the chance.

January 21st, 2012. The last day that Ashley and I went tracking together before she went home, we decided to check the bridges for scat to see if we could find any lion sign. The bridge to the south of the foothill, in zone 3 should be marked by Sneaker, if he's here. This is where we went first. It didn't look like anyone had driven this road in a while. When we started walking and checked the ends of the bridge and saw nothing but old scat from prior markings. When we crossed the bridge of the canal there it was, fresh scat. The first question in my mind was why is he marking the bridge itself now? Sneaker has marked the ends of the bridges for a long time. This is his pattern. The scat wasn't the usual size for Sneaker either. A key element in figuring out whose scat belongs to whom, is to become familiar with the locations of each cat and know where they mark territory. This process has enabled me to identify each lion no matter where they are in the study, even without their tracks. This scat most likely belongs to Henry. The scat looked to be a week old. Henry had moved over to the lake and to this day still hunts there on a random basis. He moved there after Hookah and her cubs took over Suka's range.

While searching the base of the foothill on the other side of the bridge, I found the carcass of what looked to be a lion in the latest stages of decomposition. The road had been graded, and the carcass was half buried yet, somewhat exposed enough after the rain to know the coat was that of a partial lion. In my mind I wanted to think it was a bobcat. However, the coat of a bobcat doesn't look like a lion's coat at all. I needed to see the head or even a quarter of its dental profile to determine which kind of cat it was, or if there was enough left of it. I don't know who else it would be. The one thing about clay soil is that it preserves whatever is buried beneath it. In October, she began venturing back over to travel a nearby mountain chain. Suka was killed on October 26th, 2011.

January 25th, 2012. It finally rained. The sky was still filled with dark purple clouds that threatened more rain. It was mid-morning. The air was cool, and slightly windy. I took a chance and drove out to a foothill where I'd seen sign of Henry. I couldn't drive off of the paved road because I'd sink. The ground was soaking wet. In some places the clay soil looked dry but, it wasn't. I've seen telephone trucks sink up to their axles from misjudging the grounds moisture content. I decided to turn around, and drive out to the bridge. Getting up the road paved with decomposed granite was scary. Thank heavens for four wheel drive.

While standing outside the truck, I looked up on the foothill. On a large boulder, I could see the unmistakable shape of a mountain lion. It sat alone in the wind looking down at the surrounding countryside. He noticed me. We watched each other for a while. Neither of us had anywhere to be. As the wind whistled through branches, I realized that I was looking at Sneaker. He turned his head to look out over the valley again, then turned back to watch me some more. It seemed that not a soul was out...except for us. After a while, he stood up, and moved flawlessly along the tops of the rocks until descending down behind them, and out of sight. What a beautiful cat he was. Aside from the fact that I'd apparently caught up with him while he was hunting, there's something magical to me, to be able to share a few moments of life with a mountain lion. No one is in a hurry, there is nothing but silence, and words are not spoken but, felt. It is unique when two different species' at the top of the food chain respect each other's distance, and yet, share the same ground.

In a few days the soil will dry out. I'll check the cameras again. This is about the same time of year when he stalks the trailer at the top of the hill. He really wants in to see what's in there. He does this every year around the same time. Sneaker is a little more than eight years old and he's still pretty healthy.

I wonder if Henry will look like his father when he's eight years old. I hope he has the same disposition. His mother was brave to raise him in new territory. Henry began as all cubs do, a happy baby with an adoring, protective mother. Who would have thought he would become a motherless cub, who would learn to survive on his own at the age of nine months? It is my hope that Henry survives his subadult years to take over the place of his father. In time…we will see.

Henry is the next oldest male in the study now. To date, he is 21 months old (8/12). He's survived on his own for 1 year. He has a beautiful red coat with faint black spots, his tail is full, and tipped in black. As a cub he survived on birds, frogs, and squirrels. He now hunts larger rodents. Soon, if he hasn't already, he will hunt the coyote that once hunted him.

I've notice that the coyotes in my pictures and videos are becoming more scares. When caught on film they are moving much quicker looking around a lot more. Other than Henry's tracks and scat, the only reason I know he's alive is because the red light on my camera scared him. A video of him in half stride was caught with his tail whipping to one side as he lept into the brush.

Learning a mountain lions habit for seasonal range change is difficult. It depends on each lion or lioness. The mountain lion Sneaker would re-enter his habitat at the same place during the same time of year to make his rounds along the same paths as he did the season before. He checked known locations for markings of a possible female in estrous. Sneaker would exit in the same direction as he did the last time. The lioness Hookah has some same patterns as well. She has her cubs in the same vicinity if not the same location as she did the prior litter. She raises them in the same places until they reach a certain age, and then takes them to a higher elevation where they have more room to roam away from the occasional human. They disperse from this location, and then return to the same location on their seasonal range changes. They retrace their habitat locations as they left them. After traveling the locations they were familiar with, looking for markers of females, they come back around to leave in the same direction from which they came. I only have one male that came back when he reached maturity at the age of three years and five months. I tracked him closely.

As females, they adopt habitat on the outskirts of their direction of dispersal. Their habitats overlap their mother's if there's room. They have their cubs in their center of home range, and then bring them to the overlapping territory of their mother's if a space is available. During this study, the spaces for overlap were available. The overlapping territories happen to be along riparian zones. Barbara Alice is what I believe to be the matriarch of the group. Though her center of home range is the Kings River bottom, her outer lying habitat for seasonal range change is here. Hookah's territory overlaps with hers to the east, Suka's habitat overlapped with Hookah's to the north east, Yinka's habitat overlaps with Hookah's to the south east. Since Suka died, Hookah's general habitat grew. Now Celia's overlap with Hookah's habitat is further south east still encompassing most of Suka's actual range.

Yinka, the possible sibling to Suka, with a minimum outline that was closely the same, has a center of home range that encompasses more agricultural land than either her mother or her sister. Though the overlap occurs at the south end of the riparian zone, it is more limited by way of corridors for travel to this area. The majority of her travel corridors consist of citrus groves, and the canal systems when empty. They provide a grassy surface with larger puddles of water where remaining fish linger, and where smaller nocturnal prey species' are abundantly available.

The locations in which they inhabit are based on hierarchy. Barbara Alice has the largest home range. Hookah, mother to the remaining mountain lions has the largest and the best habitat closest to the base of the foothills yet, extending into agricultural land equally. Her habitat has the most cover, water, corridors to travel, access to foothills, agricultural land, and prey species'. The first of the two daughters, Suka, had habitat closely resembling Hookah's yet she shared the north east overlap within the main and secondary riparian habitats. The second female, has the third habitat which is significantly different in natural riparian habitat until she reaches the overlap. She has a decrease in continual corridors but, with the same prey species' as noted along with wider spans of agricultural land and canal access.

All four lionesses have a riparian area, agricultural ground, and access to water ways and corridor's to travel through while hunting. Yinka shares an overlap of the south end of the largest riparian zone, and had the third overlap. Each habitat differs significantly in cover the deeper into the Ag land they get. It is possible that with Suka gone, she will move up to adopt her overlap. More research on this subject needs to be done. Her cub will not disperse for another year. The sex is unknown. Upon seasonal range change back to this territory a mountain lion will enter close to the same place he exited. They hunt along all of the zones but, stay longer in the zones with more prey base. The change in zones occurs when there is less prey. The mountain lions will venture through an area for a few days, hunt that property, and then venture to another area. Ultimately, they will return to their main habitat.

With Henry in mind, it has been encouraging to watch this cub survive against all of the odds he's faced. Note: On the day of the 21st, Ashley and I had found Henry's week old scat mark, then on the same day, found fresh scat from him marking territory by the lake. This was timely due to the fact that Sneaker had entered into everyone's range. Henry had moved over to the lake and to this day, still hunts there on a random basis. He moved there after Hookah and her cubs took over his mother's range.

No Date- I am still surveying for mountain lion frequency of travel with fixed points, to maintain knowledge of mountain lion survival in our area. It has been about a month since I've gone out to look for the cats. I wasn't sure if I'd even find any. I checked all fixed points, and found no traces. Since Ben had gone back to the river, I thought I might find Hookah and the cubs again. I haven't seen evidence of her, not even through my camera lens in a month. Others have seen her with her cubs in the mornings. She is said to be watching over them as they play in the grasses.

Something familiar did happen at the lake today. I had pulled over to walk back into the citrus grove where I'd found Henry's scat in January. It was a warm, beautiful day. The red winged black birds were everywhere. When I got to the approximate location where Ashley, and I'd found the last scat, I stopped to look out over the lake at the different shades of green. I thought about how lucky we are to have this riparian habitat, and all of the wildlife that lives here.

123

The lake bottom is overgrown with reeds leading the human brain to think that shallow water fills the lake bottom. Actually, there is a channel of water that winds through the bottom land. Its depth and sinuosity is hidden by the reeds. Nature's sounds of bird songs and sedges brushing together were interrupted by the sound of a large bodied animal entering the water to make its way across the channel, in my direction. When the sound of the water's wake grew silent I knew the animal had reached the bank closest to me. Had this been the buck, the rack on his head would've shown above the grasses. I saw no movement in the vegetation and yet, a minute later all of the red winged black birds flew up into the air. The animal was moving through the reeds in silence. I knew who it was. Henry, the little cub without a mother, had become a mountain lion. If time allows, his shadows will be cast by future generations.

Chapter XX
Study Criteria

California Mountain Lion Study Criteria

A personal study done to better understand the behavioral traits of the California Mountain Lion, and the impact of obstacles they face with increasing human population, within the lower foothill region, bordering on ranching and agricultural land with rigorous farming operations.

The goals of my first study include:

- Designate a study area by fixed points.
- Estimate the mountain lion population within the study area.
- Find each lion's home range, and track seasonal movements.
- Follow each lioness from mating cycle, birth of cubs, attain count of cubs at birth, cub rearing, monitor mortality of cubs, and number of cubs from each lioness at time of dispersal, and return of adult mountain lions on seasonal range change.
- Evaluate and monitor the overall health of lion's by sight and hunting patterns.
- Observe the lion's prey availability during each season, and preferential prey.
- Investigate lion-human encounters, and deduction of non-confrontational/confrontational lion behavior with personal and eyewitness accounts.
- Identify lion's activity patterns in relation to habitat most often used, and the distances between local homes, and agricultural activities.
- Develop questions pertaining to current agricultural practices, interview farmers or foremen.
- Compare interview results with track and sign results before and after seasonal farming operations.

The outcome from each goal will give important information to help in evaluating not only the number in mountain lion population within the designated study area, but how they stay so elusive while adapting to agricultural operations, and living in such close proximity to human residents.

After proof of mountain lions is established by tracking, an amount of acreage will be chosen to survey. The study area will be checked for frequency of travel, and such places will be marked as designated fixed points. The area will be reduced into actual travel zones with mountain lion travel recorded in miles. The area recorded will become the new study. The identification of each mountain lion will be distinguished through sight, track and, or sign.

Fixed points will be marked by GPS to map which lion is using the area most frequently and which lion's territory is overlapping the other. If possible, the center of each lion's home range will be identified but, not documented in this publication for the sake of their safety. Their daily or weekly activities will be noted with the length of their stay, and time of seasonal range change.

With the identification of each lion or lioness, the area in which they spend the most time will be mapped. The mating cycle for each lioness will be identified by tracks. Two mountain lions of mature age, in the same location, either observed by myself or residents will be noted. Separation of the lions by track evidence will help to give an approximate date for the birth of the cubs. After cubs are of an age to travel with lioness, the tracks may be observed at fixed points for a cub count. Monitoring mortality of cubs at the beginning will be difficult if no denning site is found. The observation of the traveling cubs number decreasing will be the only mortality sign that can be observed. The number of cubs at time of dispersal will account for the survivals rate. The travel patterns of a lioness and cubs will show an approximate area with possible direction for which the subadults will disperse to. This area will also dictate where the mature adults will return to for first seasonal range change.

Considering the physical health of the mountain lion community living within proximity to local residents is pertinent due to the fact that less confrontational encounters should arise if the lions are hunting and catching prey regularly. Drought can be a negative factor with a lion's sustainability. By documenting different wildlife prey species' availability throughout the lion's seasonal habitat in each zone, a preferential prey may be established along with a prey substitution, and which season that substitution comes into play. Journaling and mapping found kill sites will aid in this research.

The investigation of mountain lion encounters from residents is valuable in assessing the type and location of encounters. Determining whether the human activity was provocative to the lion may help establish why an encounter was either perceived as confrontational or not. Both subject's actions and reactions can be linked to behavior. To which party the confrontation could be so perceived, is valuable for the health of not only the residential community awareness but, for the health and safety of the resident mountain lions, and their cubs. These factors are important for sustaining a healthy Eco-system.

Identifying each lion's activity patterns of hunting, resting, denning, mating, and drinking areas, the determination of distances between their locations of natural habitat, Ag land use, and distance between residential homes can be averaged. The outcome will show the average distance of homes apart with the most frequent lion travel.

Determining the travel patterns of each lion through agricultural lands will help to show a time lapse in their travel routines when agricultural operations are in effect with these elements to consider: (1) Determining if chemicals or non-chemicals are used, and how often, will give insight to the numbers, and types of species' present in each habitat study area, and how long the break in mountain lion travel may occur; (2) The known amount of time spent in the field by workers will aid in assessing the impact on wildlife from this activity, and the break in travel routine in regards to time element before resuming natural travel.

The interview questions developed for farmers or their foreman have been tailored to fit the types of farm operations for crops in production. Chemicals or non-chemicals either sprayed or applied, will further the understanding on the mountain lions break in frequency of travel for certain routes during farming operations. The information about the amount of human exposure to wildlife from farming operations will help to show a possible pattern to how mountain lions accommodate agricultural practices into their daily travel routines.

Chapter XXI
Reflections on the Study

The behavioral insight documentation on the mountain lions living within their own habitat, at the lower foothill region of the San Joaquin Valley, was taken from my field notes exclusively in an effort to understand mountain lion behavior first hand. With my documentation and eyewitness accounts of rural residents that live and work within each of the study zones, non-confrontational behavior has been observed throughout. This behavior has thus occurred within close proximity to human exposure. The insight gained as to why mountain lions, also called cougars, have chosen their niches, should lead to the behavioral understanding of the cats, and the realization that they have not demonstrated any significant threat to their human neighbors within my study.

There are several sources for education on a keystone species', what they are, and do for the environment, and their importance. To put this into context regarding the Puma Concolor; there are usually less larger carnivores in a given area as opposed to other wildlife species'. Large carnivores eat smaller or weaker carnivores, omnivores and herbivores. They keep all of the species' numbers in balance. No matter where I look for this reference, the terms are all indicative to one another. I have found that there are normally fewer mountain lion predators than other species' in any given location of my study.

I have recently learned that mountain lions are an apex and keystone predator species' within some ecosystems. There is some controversy in regards to where a mountain lion is regionally regarded to be either an apex predator species' or a keystone species'. Here in the valley, I consider the mountain lion to be an Apex species' because it has no predators other than man, and "Man," in California, is forbidden to hunt the California Mountain Lion.

According to Harley G. Shaw (2007) in, "The Puma Field Guide," deer are the staple prey of Puma throughout North America. Shaw also wrote, "Does, fawns and bucks are taken approximately in proportion to their occurrence during late winter and spring (Shaw, 23)." After looking at my own research I realized that while I saw what Shaw meant by "late winter and spring," I also saw what he meant about, "their taking has to do with occurrence." Here they occur in the late summer as well. It makes sense to me that there would be a time of year when deer are primarily preyed upon. Deer are not usually down here in the valley unless they are at the river. The river is seven and a half miles from Wahtoke Lake. One herd of mule deer comes down by following the river. The deer that end up here are chased down by one or two mountain lions. While the deer feed on cattle grazing land in the foothills in the winter, Sneaker would find them, driving them down into the farm land. There have been occasions where he's been seen with a mate while running them down. Years ago, there were so many deer here that the established migration pattern was passed down to future generations.

The deer I have obtained pictures of in this small rural area are seen mostly in mid-winter, spring, and in the summer. They come down to this lower elevation of 420 feet, at the base of the foothills to feed on grass in cattle pastures and, figs in riparian areas where fig trees grow wild. It has been my observation the deer have used the over grown fig trees for shade cover, resting place, and a food source. I have noted their presence in the early mornings, and late afternoons in one specific riparian area. The deer herds have been as large as approximately 20 plus in number to as small as three. In mid to late January when the short sprigs of grass begin to grow, the deer start arriving. By the time the grass has a growth spurt, it is easy to see how large or small the seasonal deer population is because when bedding down they later leave flattened down grasses. The track and sign to indicate their numbers have been documented, and pictures from an infrared camera show the deer's presence.

A small Mule Deer doe found my camera when she was foraging under her favorite fig tree in late summer (2008).

Mountain lion scat found marking a small riparian route boundary.

While growing up in our small rural community, I had always heard Mr. "W" say, that even though the mountain lions here were few they kept the rodent population down. As a past Silviculturalist for the U.S. Forest Service, Mr. "W" has seen many mountain lions throughout his time. During my research of this species', I have observed the tracks of lions stalking birds, rabbits, raccoons, opossums, squirrels and various other prey species' like deer, wild pigs, coyotes, and the occasional domestic cat. By examining left over bones from carcasses at kill spots primarily of three lions, I found their prey base to be deer, pigs, rodents, and birds. My thoughts on what these cats were eating was backed up, when I found their scat. Either shards of thick bone or small bones and hair, along with teeth from the prey were within the excrements, and also at kill sites.

Chapter XXII
Mountain Lion Estimation

The estimation of mountain lion population was accomplished by tracking in each zone, and observation. Throughout a five year time period, three predominant lions were noted with the exception of one gray male, transient, who was secondary in keeping the gene pool moving. During the last year, an increase in population by one lioness and possible transient lioness was noted. The total for mature lions that use or have used this basin was six.

One dominant lioness reached maturity and had two litters of cubs, possibly three. The total number for her known cubs is five. There is some mystery still as to whether or not Suka was Hookah's first cub. Suka, one of the three lionesses, gave birth to one cub. The total number of cubs born to the study is six. The loss of two cub's lives decreases the offspring count to two current lion cubs, one subadult, and one anonymous cub in the study. As of 10/08/12, there are no subadults left in the study. The two cubs grew into subadults. Clause and Celia are not traveling together any longer. They have moved on. Two mountain lions are deceased leaving three lionesses possibly and Barbara Alice to travel in and out of their ranges. The gray male has not been seen since last year in the spring of 2011.

There were three young males born to the study. They are at different stages of maturity. Ben, the oldest of three years and three months (8/12), has reached maturity and made one seasonal rotation between two habitat areas. He spends the majority of his time away from where he grew up. His range is larger than Henry's range which only encompasses part of the original delineation of the study. Henry is twenty one months old (8/12). He will reach the age of maturity for breeding around December of 2013. With Sneaker now deceased, Henry will be able to span outward, and adopt more ground for his home range. The youngest male that now resides to the south east of the study is Clause. He is eighteen months old and still has his big orange spots (8/12).

These three cougars will have overlapping territory due to the way their home ranges sit right now. It may change later. Henry is at the center, Ben's home range is to the west, and Clause's range is to the south east. Ben has already discovered Henry's boundary marks. In time the two younger males will spread their ranges out to encompass more land even if the majority of it is on the valley floor within crop land. With Henry's presence, Hookah has been pushed farther into the agricultural land where she's been hunting the canals. The pooled water holds aquatic life, and wildlife resides close to these areas making it accessible for her to hunt. The thick, lush water grass that covers the majority of the canal bottom makes it possible for her to hide her scent while she's in her gestational period.

June of 2012, was the last time Wahtoke Lake was full. Due to drought in the Central San Joaquin Valley, the lake wasn't able to be used for water storage. In 2013, and 2014, a small amount of water was allowed to runoff into the lake bottom from the canal. When the lake is full, it replenishes our ground water which in turn helps our water wells.

Wahtoke Lake is a small natural lake, and home to a diverse species'
population (2013).

Chapter XXIII
Overview- Prey & General Habitat

Zone 1 of the study was where Hookah had two known litters. It is also where they spent much of their time during the first few months. Within this section the habitat is primarily riparian due to the fact that she had to be close to water and small prey. Zone 1A is an amendment because one lioness changed her territory to encompass the empty canal system which not only provided prey but secrecy in day light hours. She did this after she was put into hide mode from either a trailing rancher, or the seasonal presence of Sneaker.

Zone 2 has more than one section that the mountain lions roam. In one specific area people and mountain lions share the same paths at different times of the day. People walk and jog on the canal banks. A number of residents ride horses around the perimeter of unfarmable land, field workers tractor, and others work on foot in the early mornings. Most people's travel starts around the break of dawn. This is when the lionesses with their cubs have been seen the most. Neither of the lionesses has been aggressive. They just go about their travels seeking cover, and prey with their cubs in tow. The lion families are observed passing through by the field workers.

Hookah is not the only lioness that has used this place as a safe haven. Suka has primarily raised Henry here. This is part of her overlapping territory with Hookah. The prey species' and vegetation are the same where she came from, but more abundant here. This habitat has thick corridors, and it's near water. The deer population, in certain months, and number of wild pigs increase the possibility of hunting enough prey for a mother to support her growing cubs. Barbara Alice and Sneaker pass through this zone as well.

On the top of the foothill in zone 2, the limited access to humans remains an advantage for Hookah. This is cattle grazing land that people normally have no access to. This area is only traveled by the rancher who owns it, or his cowboys. A single dirt road leads to the top of the mountain. There are two ponds on this mountain, the first is seasonal and the other is fed by a natural spring. There are rocky outcroppings for catching some sun, or getting a full range view of wildlife, and the valley floor below.

This elevation ensures that there should be no human surprises.

When both of Hookah's litters were approximately four to five months old, she took them to higher elevation of 1070 feet, to finish raising them. This move can also be attributed to seasonal range change. The colder air on the valley floor, and the less than adequate prey availability during the winter months for a lioness feeding her young, is indicative of a lionesses range change. In late winter/early spring, Hookah will travel to hunt deer in a different section of zone 2. On warmer days Hookah and her cubs' tracks can be seen when taking a walk to the adjoining riparian area where easy prey of bullfrogs and ducks for the cubs is available. A single lion like Sneaker, has more chance of supporting himself on the percentage of rodent prey available during the fall and winter months at lower elevations.

There are several sections in zone 2 that mountain lions travel. The slough that connects the two watery regions holds water because there is a leak in the canal system. The wildlife of: ducks, bullfrogs, coyote, pigs, raccoon, opossum, ground nesting birds and hawks, Great Grey Heron and Snowy Egret, owls, bobcats, a beaver family, gophers, squirrels, snakes, gray fox, and the mountain lion, all frequent this area. The major riparian area of the Wahtoke Lake bottom houses all of the same species', and at times, mule deer.

The dendrology of the riparian area within zone 2 is thick with reeds, cat tails, water grasses, Western Sycamore, Valley Oak, Willows, Eucalyptus, Fremont Cottonwood, invasive Mesquite, and volunteer olive trees. This landscape provides a more dense cover than that of other zones. Obscured by thick foliated corridors along the water ways throughout zone 2 the mountain lions sporadic presence remains undetected. The difference between the two riparian sections is the wild, non-indigenous, mesquite trees that have all but taken over the south end of the lake.

The foothill sections of zone 2 have a dryer climate with different dendrology. The brushy caves are the perfect place for a cub to play, and rest. Just as people view lions to be a threat to human children, a lioness views people a threat to her cubs. Hiding the cubs is easier in the natural rock caves obscured by scrub live oak. The brush is drought resistant and grows in soil types that are well drained with minimal organic content. The majority of this growths exposure is on the top of the mountain leading down on the northern slope about 20%. The scrub live oak grows close to the rocks and following the ravine where seasonal water flows. Cottonwoods and Poplar trees follow the water runoff from the spring.

Zone 2 and 3 are separated by bridges. The foothill with a lower elevation of 605 feet, and the neighboring pasture land is at the elevation of 538 feet. Even though Suka shared zone 2 with Hookah, I had first discovered her and Sneaker, here. The available cover is from citrus trees. Drip irrigation provides a water source and helps to keep the warm spring afternoons and summer evenings comfortable. The elevated foothill exposures are northwest and northeast.

The eye witness accounts obtained from this study provide more than enough information to show that the lion population that inhabits our small community, shows little to no aggression to humans. These lions have succeeded in coexisting with people, and trying to maintain anonymity to live their lives in a society where people have taken over much of their habitat for housing or agricultural practices.

The sharing of smaller home ranges often indicates a larger population base. I've seen from the girls that people who've observed them think that there are too many lions in one small area when the girls have cubs. People don't realize that the cubs will leave and the mom's, if not a resident the majority of time, will also leave after her cubs go off on their own.

My thought on this subject is that: like people choosing a residential area based on cultural, economic, and social needs, the lion does the same with regards to choosing habitat. In essence, the economic needs for a lion are water and prey. The cultural aspect would relate to cover for raising young, and social needs would be based on proximity for mating cycles. For instance, Las Vegas, Nevada or Los Angeles, California: People have found an area where they can live with the natural resources available. The amount of resources available determines the degree of population. Scarce resource's means small population, and more resources equal a larger population.

Another example is cattle: In the foothills there are areas that have more mountain lions than others. When a cowboy goes to round up his range cattle and they are not where they normally can be found, he may head to a shaded draw to find them. The draw provides cover, sometimes water, food and a temperature variation that makes the location more comfortable. To a given degree, the cattle are all within proximity of each other. They have found an inhabitable location where their natural resources are available to sustain them. Therefore, the existence of mountain lions is no different except the lion's territory is much larger.

The Desert Cottontail (Sylvilagus audubonii) is one of the most abundant prey species for the mountain lion, bobcat, and other omnivores in the area.

The "Specially Protected Gray Fox of California," The Gray Fox (Urocyon cinereoargenteus), is omnivorous and hunts alone (Elevation: 400 feet).

The Common Great Horned Owl in California (Bubo virginianus virginianus) sitting in a cotton wood tree. The owl sees everything.

The Raccoon (Procyon lotor) is visiting the lake in search of dinner.

Chapter XXIV
Travel & Preferential Habitat

After delineating a study area which encompassed a large area of acreage, fixed points for frequency of travel were established. The study was divided up into three primary zones. They are referred to as zones 1, 2, and 3, with one section amended as zone 1A. The reason why the study was divided into sections and not kept as a whole is because some residential homes with dogs were too close in proximity to the creek bottom which is used as a connecting thoroughfare to different habitable areas. Frequency of travel was established by track, sign, and visuals. There are several points within each of the zones that show frequency of travel on a daily basis by track, some more frequented than others, and the directions traveled to, and from are visible.

Record keeping for frequency of travel for each lion, lioness, and cub(s) was done by GPS mapping with waypoint markers to establish points on maps. Markers were used to plot routes for each lion, date, and label, use specific color, and print out to store. After the breakdown of area using fixed points, the original study area was decreased in size due to actual lion travel. The outer lying areas are considered to be incidental since a lion would only use it for hunting on a sporadic basis. Checking fixed points twice a day was beneficial for establishing the mountain lions daily or nocturnal hours of travel. The details of their activities such as the directions from which they exited, and or re-entered the habitat, and area covered were monitored. Wildlife and domestic animals were also used as indicators for the nocturnal activity patterns of the mountain lion.

The following is a description of preferential habitat, and of mileage in travel patterns from each lion. The original study circumference began as 19.08 miles but, decreased according to fixed points from lion travel. The area was not recalculated to show a smaller range because of nocturnal hunting in areas mostly unavailable for my travel yet, not for a lion trailing coyotes. Later expansion of the original research area was necessary due to a lioness incorporating more land into her range, and the need to survey that new territory. The inclusion of new territory was then recalculated to 23.06 miles in circumference or the equivalent of 7,212 acres. The determination of frequency of travel at fixed points for Hookah brings her mileage to 28.58 miles of meandering paths within the study. Suka's range was smaller as she had spent less time here. The mileage for her routes within the overlapping territory amounts to 14.97 miles. This mileage takes into account the distance traveled from her last habitat into zone 3, 2 and 1A. Considering that she traveled her routes more frequently than Hookah, because her range was smaller, her amount of travel was more frequent. Suka's mileage is conservative before adding her routes from the other foothill mountain chain. Though I know her prior area of travel, her travel patterns in that range are unknown.

Two of the lionesses travel close to the same routes but, at different times of day or night. Sneaker's mileage is less within the study because he doesn't stay as long as the girls do. His route measures to be 23.22 miles. The actual distance traveled between the three most predominant lions is approximately 66.77 miles within the 23.06 mile circumference. This number does not take into consideration how many times they traveled their routes. Hookah had the most mileage for delineated paths because she's lived here the longest, and this is her center of home range.

Following Barbara Alice's route is difficult. I have to go not only by track or sign for her frequency of travel, but sightings as well. While doing this, I also figured in the deer herd routes. Barbara's total for distance is 67.61 miles, where 30.04 miles were inside the study and 37.21 miles were out of the study zones. The calculations are not exact because there is land that I couldn't, and didn't incorporate into the mileage.

The, "Barbara Alice" finds refuge in the more densely covered habitat with canopy cover at 75% to almost 100% in places of the river bottom. The habitat she chooses to relocate to is also within riparian areas with silty soil. The animal frequents zones 1 and 2 after its range change. The canopy cover is approximately 90% to 100% which is similar to her Kings River habitat. She has been known to roam the open ground covered by tall grasses that coincide with shrub, and tree lined creek beds. The distance covered is not limited to the travel pattern of thirty miles within the study. Sightings of the animal are reported sporadically throughout the year but mostly when she travels during seasonal range change. People that live along her travel route, do not know what type of species' this is. More data on this large felid is needed. In regards to routes traveled, she crosses 4, and possibly 6 busy roads to get to the base of the foothills. I have found that lions prefer to den in familiar places. These places are established for safety, and available resources. The Barbara cat prefers the areas with more brush for travel and finding unused shelters, burn piles, and rock caves to sleep in. As with all of the lions in the study the animal keeps moving, but more often, and in a wider span. This animal follows the waterways like so many others, and has been seen crossing crop land. I have found her scat in zone 1, and tracks in zone 2. The areas where she spends most of her time hunting in the summer are both surrounded by trees. Both areas can be considered to be an oasis in the middle of agricultural progress. This is a wild felid that is really hard to get a picture of. Exactly what Barbara is continues to be a mystery, but, I will refer to her as a mountain lion since she fits into the family of large wild felids.

The canopy cover surrounding zone 1 is approximately 95% to 100%. The center of the slough is clear of trees with cover of reeds, and cat tails. The area measures 0.82 of a mile in circumference. The soil moisture content for zone 1 is focused in the center of the acreage and normally high in spring, summer, and moderate in the fall. The olives in zone 1 are not irrigated and the pond is centered while 2/3 full from irrigation runoff.

The differences in the two habitats are the olives are irrigated around the perimeter in zone 2, near the slough. The canopy cover in zone 2 where she hunts and rests is the same with the exception of the open field in the center. The slough habitat in zone 2 is 1.59 miles in circumference. The water holding capacity of the slough bottom is approximately 1/4 full with some fresh water running year around, from canal leakage. The olive trees in this zone are focused around the perimeter, and irrigated. She is able to walk or sit in these grasses to hunt without being seen. There are citrus trees to the south which provide added cover. Zone 2 is where the unplanted field in the center is wet, and green in the spring, and either dry or tractored in the summer.

The dominant female, Hookah, is in all ways mountain lion and benefits the most by her home range niche. She has not only her immediate home range of the largest riparian area connected to a secluded mountain top. She spends 70% of the year in zones 1, 1a, and 2, when she has small cubs. When her cubs get bigger, her travel pattern expands into zone 3 for hunting. This area also has more room for her cubs to grow, and go undetected by humans.

Upon determining the time frame of the co-dominant lioness, Suka and cub Henry's inhabitants, by frequency of travel, and sight until her death; she spent roughly 60% of time within zones 2, and 3. Her remaining time was spent to the east, on a foothill chain that was later incorporated into the study. The area was not open to me for research in the beginning. The fact that Suka eventually included new area into her habitat shows that while they have a specific area most visited and often called, "The center of the home range," they cross over into another's territory to encompass a wider range for the purpose of sustainability. Obvious reasons for Suka moving closer to the study area could be attributed to overcrowding by another mountain lions overlapping territory, crowding from humans, or a male making his rounds while she had a cub. My thought is that she, like Hookah, and Barbara Alice, chose another area to finish raising their cubs for the purpose of dispersal direction.

The specific area in distance covered by the cougars is larger than the actual study in square miles. Their individual path measures show a trend in travel as three to four days spent hunting, and resting in a listed zone, then traveling to another. All zones connected by a creek or slough are traveled more frequently by the lionesses and their cubs, each at different intervals. I have found that the routes the girls travel overlap in some places. In others, their routes pass within yards of one another's. They choose to walk on opposite sides of bridges. Where Hookah will travel to the slough in zone 2 by way of the pecan grove, Suka and Henry will be leaving in the opposite direction on the other side of the canal bank. They've already hunted the area over night, and into the early morning. It was time for Hookah and her cubs to go out to see what they could hunt. There is also another place where I have noted the girls passing paths. While one lioness will travel a dirt road next to an open thistle field, the other will travel just yards away, down off of the canal road into a canal ditch. The ditch alongside the canal handles runoff, and provides small game for hunting on her way to another shared location.

Suka was last seen by a pecan grove worker. She was apparently hit by a vehicle while crossing the road on her return to zone 3. Her cub who is now mature, still occupies his mother's home range as track and sign have indicated.

With the death of Suka, the dominant lioness adopted her territory and has been seen hunting for small prey by four cyclists while mountain biking in the spring. Both times they saw her; she bolted for the security of cover. I wanted to go to this area and set cameras for pictures of her. I chose not to because I feel that it would be selfish and those earlier chance encounters were enough for her. I wanted to preserve what had been infringed upon...her anonymity. Hookah had found a place that was secluded from people and had all of the resources that she needed to finish raising her cubs. The fine line of wild and familiar has been crossed by humans to mountain lions and in my opinion, not the other way around. The cyclists knew she was there and yet, still returned to that area to ride, even though I have expressed concern for the lion's well-being. I also think that when the general population shows a set pattern of travel including time, in a known lion habitat, it sets the lion up for failure. Future encounters could prove to be dangerous. This is the fault of the human. She is a beautiful cat and I love to see her, but only on the terms of inconsistency.

The dominant mountain lion, Sneaker, found the activity of farm workers and their noisy equipment amusing. He had been witnessed by farm laborers and contractors, sitting or lying on a lower foothill watching, and listening until he became subdued by the general chatter of people until he fell asleep in the sun. He made his rounds at the same time every year. He spent approximately two to three months in the winter, and returned in late summer or early fall for another visit. He spent approximately 40% of his time in this range. This male lion survived primarily on small game.

The gray lion is a lion that I have only seen once or twice from a long distance. His information comes from several eye witness accounts. This lion shared the rural farm land in the past, and kept the gene pool moving. It is possible that he visits here in the fall but, the last sighting of him was over a year ago. He was traveling through nearby property to the pond area in back of our home. There is fresh water during spring, and summer. If there is water remaining, the pond turns to a marsh at one end. Coyotes den at the other. Olive trees surround the riparian habitat where year

144

around prey of raccoons, opossum, mice, gophers, squirrels, coyotes, and water fowl are present. Except for scattered visits from ranch foremen during the summer months, there is little travel by humans in this area. Human activity picks up with farming operations in the fall, and winter. Citrus grown on the borders of the pond becomes ready for picking in November, and December. The majority of fresh lion fecal matter marking boundaries near the pond is found in the spring, summer, and early fall.

Interestingly enough, when I watch my domestic cats I see them hiding in the tall grasses like the lions of the Serengeti. They are either hunting or resting under natural cover in secrecy. During my study, I found that the mountain lion does the same thing. I had to question how they get from one location to another in daylight hours without being detected. I found the answer when looking at the base of a mountain, at the span of crop lands that sporadically take off from the base of the foothills. There is acreage within the study that has been left vacant for crop rotation. The land is densely overgrown with tall dried weeds and at times, an undergrowth of green grasses. Around the perimeters of the crop rotation plots, I have found the tracks of pigs, coyotes, rabbit, and other rodents. The crop rotation sites provide mountain lions the corridor for travel to nearby riparian zones. While traveling through this cover, rodent prey is inevitable.

Chapter XXV
Diet

Each lion's preferential diet is different than the next. The common order of prey species' is as follows: Deer in late winter to early spring, and sometimes they are available in June. Then, the wild range goats after the deer become depleted, mostly pigs, the occasional calf in winter with consideration to male or a female mountain lion with cubs, drought, ducks and water fowl, rodents of raccoon, opossum, rabbit, gray fox, beaver if they can find one, and ground nesting birds. In late summer to early fall, more deer are chased down from the foothill region. This occurs mostly by the three resident cats, than the deer hunters, and the cycle starts over again. Observations serve that a lioness with cubs will exhaust all prey in an area before taking calves or sheep. It seems as if they know it will lead to trouble. The taking of these two crops will alert ranchers to their presence making the habitat chosen to raise young unsafe.

It was my experience through tracking these lion's that the majority of kills were wild pigs that roam in large numbers from the foothills down to the valley floor within the agricultural farm land. The pigs are abundant, and easier to catch.

There were three definite deer kills that I witnessed through track, and sign. The two lionesses, Hookah and Suka, hunted in the same area to catch them. The deer and pigs feed on figs in the lake bottom. The trees also give the deer a false sense of security which makes them easier to hunt, in the late summer evenings. A farmer who lives at the base of one of these foothills told me that in the late 40's and 50's, there were large deer herds that always came down for the winter to take refuge on pasture land at the base of the foothills. Some were on their way to the Kings River. There were so many, that apparently it became an issue with agricultural crop gains.

The slough in zone 2 along the canal is home to not only pigs, but beaver, and raccoon. There was more evidence of raccoon kill sites than others. This is also substantiated by the number of raccoon tracks verses other wildlife. I found raccoon teeth in mountain lion scat. Some squirrel tails were found in the brush. I saw this on the Sierra National Forest in 2011. A lion will strip the tail off of a squirrel, and leave it behind. The fact that the tail I'd just found was not far from a huge pile of fresh scat from a mature mountain lion helped me figure this one out. Bobcats are prone to do this too. In order not to mistake one for the other, I look for other signs that can lead to more information. Tracks are a good sign.

The dietary subject gives way to comparisons in meat textures, what the lions prefer, and when the delicacies are in season. Wild large game such as deer is considered, prime," but when that is out of season, choice," is the next best thing. The texture in meat changes to goats and wild pigs, then down the chain thereafter according to relevant abundance; i.e., deer, goats, pigs, coyotes or dogs, raccoons, water fowl, rabbits, squirrels, opossum, house cats. Mountain lions associate huntable areas with available food sources. Prey substitution is a sign of economic decline. If more ranchers would keep a wild goat herd on the same property with their cattle to detour the lions from eating their calves, life between the mountain lion, and rancher could be better.

In this study, the prominent male was Sneaker. His favorite prey was wild pigs, raccoons, ducks, and bullfrogs. I wondered if his prey preference was linked to the deterioration of his teeth. In part, it could've been. I also took into consideration the seasons he meandered through during the year. When Sneaker came in the early part of the year, more prey was available than at the end of the year in fall. In fall, the prey that he was eating was what was available the most. This is another example of prey substitution.

The predecessors of Barbara Alice are known to be rodent hunters. As Barbara grew, she developed a taste for beef. Though she too, hunts rodents, this next to deer is her favorite meal. She also prevails in dominating all of the study zones when she is here for her stay. This Barbara is the animal with color. With the color that she shows, and the fact that she is too big to be a bobcat, somewhat shorter than an average mountain lion, the length of their tails being less than two thirds in length from a mountain lions, the feet being smaller than a lions yet, larger than a bobcats, the fact that she and others like her prefer caves, rock piles, barns, and brush piles, lends to the outcome that through the large possibility of human error, a new species' has been evolving under our noses since the early 1970's.

The dominant mountain lioness, Hookah, has a preferential diet of pigs and small game such as ground nesting birds, guinea hens, and rabbits. Then the occasional ducks, squirrels, and bullfrogs seem to fill in the gaps. When deer is on the menu, I'm quite sure she's ready for it. She is a happy cat and a devoted mother to her cubs. She is never far from them. Hookah has been the most relaxed yet careful lioness to watch. Her first male offspring, Ben, has been a regular male with his inquisitive nature.

Ben is a cautious cat with his human encounters because he learned that it can be painful, at the early age of 16 months. I had hoped this would serve him and so far it has. Visuals of Ben have been made by various people in a local area when they disturbed his bird hunting. He has taken after his mother in this instance. There are plenty of show heifers around for him to eat, but he avoids them. This pleases me to no avail. I hope it continues this way for as long as possible. Ben is three years old now (4/20/2012). He resides in an area not associated with this study, but I have the opportunity to hear of his presence on a daily basis. So far he's got a pretty good life worked out for himself.

In the summer of 2011, Suka's dietary preference was Rhode Island Red chickens that were stored in bulk (or so it seemed to her) from a nearby neighbor. She was short in height and slight with regard to her young age and previous nursing. I tracked her and found that she loved the chase of pigs and bullfrogs the most. It was April 20th of 2011, when I first saw her. One year ago today, I stood in front of a mountain lion that was just as amused with me as I was with her. The fact that she allowed her single cub to be present at the site where we had previously met that day, surprised me. I know she wasn't far when we set out a camera, but the nature of the curious cat was to watch silently while we chatted and worked. Suka must have been taking it all in. Perhaps she had never encountered a human so close before. I will always remember the two stripes on either sides of her eyes. They gave the appearance of smile wrinkles that an older woman would have. As she squinted in the sun, I looked at those lines, and thought that she would always look like a happy cat.

Suka unavoidably left her offspring, Henry, to grow up alone. I have lost track of him but the last time I saw what he had been surviving on, it was squirrels and most likely small rodents. I had found scat and track from him in January of 2012. I was happy to see that he had progressed in age. To date, Henry is twenty four months old (1/2013).

Chapter XXVI
Details

With the unknown location(s) of Barbara in mind, I use my truck to travel the access roads while tracking. When tracking on foot, I try to make sure that I don't get caught up in the fact that I've found lion tracks and travel to far from my vehicle. There are some parts of zones 1 and 2 that have overhanging trees and dense foliage. If I am correct in my theory about how Barbara hunts her prey primarily by ambush, I have to be really careful. Besides, it is mostly the two legged species that will produce the most danger. Therefore, I try not to go alone, and when possible I make contact with the land owner or agent, to let someone know where I will be. This research project started from taking a walk and the sightings of resident cats. I do not set out to specifically find any lion. This survey study was meant to observe mountain lions, their habits, and impacts they face, through track and sign. The times of day that I track are not within the early mornings or late afternoons at dusk. I never track at the same time of day as the last, as to not create a scheduled event. If I see one, it is a chance encounter, and treated with extreme caution. I might add that it is difficult to get pictures of a mountain lion with a wildlife camera. Placing one at a kill sight would most likely get better results but, this is something I haven't done. I do not want any lion to catch my scent. In looking back, I've realized that with every chance encounter, I was either far enough away or up wind from them. I want to try and keep it this way.

As with all farmed acreage, there are access roads. The agricultural plots that are in a clay based soil have decomposed granite for road base material. Alongside the roads is a clay loam soil type that is beneficial for tracking. It is easy to see if any lions have been in these places during winter, spring and early fall because of the wet soil. For tracking during the summer, finding tracks is more difficult and depends on soil type. Sometimes the aid of a water spray bottle helps, too. I use this limited technique when I'm not sure about the foot pad. I might spray it with a little water to see if the track shows anything different. Sometimes, it does. I don't like being wrong about a track. It's better to make sure the track was made by a lion than to assume it was a coyote.

Assuming is not a safe bet. A lions leading toe should help in reasoning if I can't see the whole foot pad. Some lion tracks have pointy outside toes and center toes that look almost even in dry soil. If I really can't tell which track it is, I have to keep looking until I find a better one to look at.

For distance between residential homes on the west side of the lake, there is a 226.82 yard average distance. There is a 271 yard average distance between five residential homes on the east side of the lake which is most widely traveled. The distance between three homes bordering the crop rotation land where the study lions actively cross from the largest mountain in elevation to the adjoining territories is measured as .45th of a mile. The lake bottom is a fixed point marking frequency of travel for lions using the lake for hunting grounds, resting, and travel patterns. They are known to cross the road from the lake to gain access to the citrus groves for cover, where they most often follow the banks of the first ephemeral creek. On average, the houses don't have to be 300 yards apart for a lion to wander through overnight. Every year, Hookah flattens down a portion of grass across the road from our house to hunt my cats at night. I always know when she's been here, even though she's really quiet now.

The length of the creek begins at point A in zone 2, and measures 2.17 miles to point B, ending at the canal system on the west side of the lake. Then the creek measure picks up again at point A in zone 2 at the back of the lake, and travels in distance of 2.75 miles to point B, and ends at the canal system to the north east side of the lake. The sinewy stream was measured using miles instead of yards because of the territory covered. The length of the canal system traveled for cover, and rodent hunting spanned from the east of point A, to the end of zone 3, ending at point B, in zone 2. In calculating the sinuosity of the second ephemeral creek, the measurement was 4.01 miles beginning in zone 3, and ending in zone 1A, that borders stream 1 in zone A. All points are recorded.

Beginning in the spring of 2011 and 2012, a wider frequency of travel span by mountain lions was noted, and continued into the fall. The rain fall had been average though the state of California was in drought, the lake was full, and the prey animals at 400 feet elevation were abundant within the riparian habitats. Even though the amount of rain may measure at average, sometimes it feels like we get more than is actually recorded. This can be attributed to the length of time between showers, temperature, and or the level of humidity. It also can be attributed to extra runoff within the watershed.

I found Hookah's patterns interesting. She will have her cubs in one area, keep them there until they get a little older, change habitat to accommodate their growth and travel for their age, and this takes into account an area of least human presence. This is the area they disperse from. When they leave, she goes off to find a mate within a couple of days. When she returns 16 to 19 days later from mating, she chooses a prior area to den in or close to the same location. The direction, for which she travels from there, is in the opposite direction of the two prior litters dispersals. She does this to avoid any random meetings with adults or subadults. Also, I have noticed that when her mating cycle is close to fall, she's well within the Ag land.

The difference in habitats lies in fewer trees, less canopy cover, and more exposure to agricultural farming operations, extended hunting periods during night to early morning hours, and fewer locations to travel with cover. She rotates habitat. During the end of September when the lake was empty, the holding pond waters were green with algae, and creeks were not running, there is less water for wildlife. When the canal water stops running, Ag wells start running. The farm irrigation through drip systems mainly occurs in the citrus trees. The only remaining water is in pools in the bottoms of secondary canal systems. Water remains in the primary canal system, yet falls to a lower level making it increasingly more difficult for her to get to.

Chapter XXVII
Drought and Survival

During the 2006-2010 drought years in the San Joaquin Valley, there was a difference in prey availability at lower elevations, as compared to good water years. The difference was that there were less deer that stayed for a period of time. The deer prefer to go to the Kings River, and some of the mountain lions follow them. Another problem is that the grass on the hillsides may begin to come up in the spring but, it dies out quickly making the wildlife search for areas with more water availability.

Heat stress on mountain lions causes them to hunt later in the afternoons, later evenings when the sun has just gone down, and into the later mornings, as opposed to mid-afternoons, and the onset of dusk. Early mornings remained the same, and each cat has its own schedule. Energy preservation is not all that is happening with this behavior. The prey species are also less active during the hottest part of the day.

When I post cameras during the hottest part of the summer, often times I get pictures of darkness. The pictures have a time, and date stamp on them. When I find these pictures I know to go back to the site and look for track or sign I may have missed when I retrieved the camera. When I find sign, I know that the photos are showing me the time of day or night when a certain lion hunts or travels through a given area. On average, I am correct. Therefore, I am not assuming that the heat changes a lions hunting pattern. However, it also depends on how hungry the lion is, and if it has a specific destination to get to in order to hunt later. Just because a photo does not contain the full image of the lion does not mean it wasn't there.

During the prior drought year of 2009, frequency of travel was limited to agricultural spreads due to higher temperatures in the triple digits, for cover and potable water. Mountain lions prefer agricultural settings where they can seek comfort in a citrus tree above the drip irrigation for a shaded resting place. The irrigation of the citrus trees also provides small game and a water source when seasonal watering holes and sloughs are dried up. The use of agricultural wells to fill irrigation ponds was essential for the lion's. They navigate between watering holes if they are not within proximity of a canal system. Each lion has a preferential diet and shares common watering holes with other lions due to overlapping territories. Some of the prey that the lionesses track within the study zones is intended for teaching their cubs to hunt. The availability of such prey during the hottest months is at a lower elevation closest to watering holes and riparian areas in early mornings, and late evenings. Some small mule deer were noted, and photographed in tall dry weeds, dry creek bottoms, and under trees where a cross breeze blew. The average time of day I saw the deer was around 8:00 p.m. in July, late summer. The herd count averaged between three to four or five to six on different days during the hot summer week. In the early spring, the deer population can climb to as many as 20 plus.

Chapter XXVIII
Farming and Pesticides

The farming operations that take place during the year are varied upon location, giving the lions alternative travel routes. As surveyed, three different farmers combined with almost twenty eight hundred acres, report the reduction in amount of pesticides being used. This reduction is due to the County Agriculture Department restrictions resulting from studies done on invasive insects. Using this data, the Department of Agriculture determines which prescription to recommend for infestations, or the prevention thereof. I have noticed more wildlife within this farming acreage as compared to twenty years ago, when more pesticides were being used. An example of wildlife that has made a comeback in terms of small numbers is the ground nesting birds; the road runner, and the meadow lark. When I was little about thirty years ago, I could see and hear meadow larks all the time. Then as pesticides were applied in stronger doses and more frequently, the populations of ground nesting birds seemed to dramatically decrease in size.

The use of wind machines and other mechanical equipment has not inhibited the mountain lions travel activity. They have become adept to the noise of running equipment and use it towards their benefit in some cases. The valley floor is colder in winter due to lower elevation. I thought that, normally mountain lions would travel to a higher elevation sooner than what I have found. However, the operation of wind machines keeps the warm air circulating; the lions have more opportunity to stay longer. This is taking into account lions with or without cubs. With the warm air circulation, rodents are more prone to be above ground. In the early mornings, prey is available and the lions without cubs are still within close proximity to water.

The farm practices within Suka's range preference would only stop her travel through an area under operation for about three to six days. This fact depends on the presence of humans, and the chatter lion's associate with them. I have found that the female lions keep a wider distance from people than males do. The only exception to this is Hookah denning near our home. Our home is approximately five hundred yards from her pond denning site.

According to an interview with a Ranch Foreman, the average time a picking crew spends on a 20 acre block of citrus, is one day. With adjoining acreage of citrus to harvest, the crew will come back one or two weeks later. They spend one more day to complete the work on the same amount of acreage.

With the degree of noise from activity on the day of harvest, and then a two week lull between harvesting, the wildlife has a chance to re-acclimatize before the next harvesting cycle of citrus. With IR cameras in place, I have documented that it takes about two, and sometimes three days for the nocturnal activity of wildlife to reconvene after a one day harvest on twenty acres.

Pictures retrieved from an infrared camera showed wildlife of raccoons, coyote, and bobcat, early in the mornings and is consistent with the tracks left at the kill sites. The young males born to the valley floor are prone to hunt the citrus groves, and vineyards. They take refuge in the trees for shade cover, cooler temperatures, prey availability, and water. They are an asset to the farmer because they keep the coyotes and pigs from chewing up the drip system hoses when looking for water during the summer. Their presence will not be seen accept for a track or two. They stay out of site and when the farm operations begin, they vacate the area.

The mountain lion is accustomed to constant travel. Avoiding farm operations during the day is not a problem because they mostly move at night. For example, Hookah hunts at night hearing where spray rigs are in operation. By early morning she's either finished, and settling down or she's still moving opposite the sound of machinery. The sound of cars or field personnel commencing their work lets her know she needs to find a different place to rest or hunt. Night operations of spraying on multiple pieces of property only detours her for at most two weeks until things have settled down. Then her journeys may bring her back to the area.

Foremen who irrigate usually have their radios on. This makes the mountain lion aware of their presence. The lion can hear a foreman's ATV or truck from a long distance away and would prefer not to be seen. Noise on the part of the foreman is in his best interest. It will help him avoid interaction between he and a surprised lion, especially when irrigating in the evening or at night. With the mountain lions presence in the Ag lands, less squirrel poison should be needed. It's important for the farmer or his agent to practice the proper distribution of squirrel poisons. While one mountain lion loves to hunt the rodents, it is only one. If an increase in squirrels occurs, the poison distributed correctly should decrease the squirrel population yet; leave some for the mountain lion.

An example for why poison should be distributed properly: Booboo the bobcat was still a little guy when I first stumbled upon him down in the pasture. I'd gone down to retrieve a camera. While navigating my way down the steep embankment I caught a glimpse of something moving. I looked to my right, and there was this shocked little bobcat just sitting there. He had a very puzzled look on his face. I was a bit shocked too. I had to get down that cliff so, I just talked to him on my way down, not looking straight at him. After grabbing the camera, and changing the SD card, I had to go back to set it out again. Back down the cliff I went and, again I met up with Booboo. We walked around each other. I set the camera, and left. I'd found him in the orange groves across the street a few times. He'd go there to hunt the squirrels every evening. The squirrels grew fat on the rotten oranges that dropped off the trees. How do people think they get cleaned up? No one goes out there to rake them up or throw them away. I've gotten some great pictures of Booboo.

One day I saw something moving in the second row as I was on my way home. I stopped and got out to see who it was. Booboo had really grown. Those squirrels had done a good job for him. Then on a different day I was on my way to town. I noticed the poison out in piles along the fence line. I knew this couldn't be good for the little bobcat. So I stopped the truck, got out, and found every pile. While I was kicking it around, I thought to myself, "This is really going to mess up his health. He's not going to have anything to eat." I didn't see him anymore after I'd found the poison set out that way. There must not have been enough for him to eat so, he moved on.

Not too long after Booboo had left, Ben came home. He also hunts squirrels under the pomegranate tree. They all do at some point. Ben really had to work hard to find squirrels anywhere over there. No wonder he needed a coyote to eat. There weren't as many four to five pound meals to catch. It would've been nice to keep an even balance so that the wildlife could survive. We do not poison the squirrels on our land. I find it funny that the farmers across the road try to kill them all off yet, all of the rotund little morsels we have over here, go across the road, gather the old oranges, and bring them back to their burrows on our side of the road. They're just cleaning up what people don't. With rotten fruit come flies, and fly larva but, not if you have enough squirrels.

I'm surprised by what I have found. In the beginning when I'd selected an area to study, I thought the mountain lions within the study zones would travel in wide circles. They don't. In fact, based on one lion and two lionesses, I found that the females meandering trails lead in opposite directions of one another, and the males make their rounds searching for an open female. I set out to find tracks of mountain lions to establish frequency of travel for the fixed points; I thought I would be looking at a much larger radius per lioness surrounding what I had selected. I didn't. I found that out of four zones, the two lionesses would hunt half of the same areas at different times. Sometimes their trails would be several yards apart. When that area had been hunted to the point that deer were nonexistent, the pig's presence was thin, and small game was low or hiding, they left to go to different zones for a few days, then ultimately meandering back to the main zone. Their habitats overlapped but, there were times more often than not when they would spend time in the center of the overlap which consisted of two zones. This place just happened to be near a body of water with extensive riparian habitat, and secluded groves of overgrown olives. There were two sections in each zone that they shared the most. The travel routes of all the lions do go outside the study boundaries. For the most part the lionesses have a concentrated presence within the designated study zones. This presence is observed on a daily basis. The borders of each zone may be slightly separated in some locations due to more human activity but, are relatively close together.

Near the study zones boundaries' the entrance and or exit trails decrease in the amount of track frequency. In seeing this I know that they spend a lot of time retracing their steps along a given radius. The routes between zones used the least are traveled when changing habitat for hunting. The ends of zones 1, and 3, are for seasonal range change.

My questions concerning "What constitutes the fringes of society," is relative to the distance of homes apart, Ag land, human activity, and what's on the other side of the road that they are trying to get to. I looked at the distance between zone 1, and the Kings River which is seven and a half miles. I used an aerial map to view each zone and each town spanning north west and south west, heading west from the foothill region. The homes bordering the outside of each small town to the South West of zone 3, have fewer homes in close proximity of each other as compared to small towns to the north west where homes are closer to each other in most cases, and forming small communities. The distance from which I measured was approximately nine miles to the south west, and nine and a half miles to the north west. It has become apparent to me that while river, and canal water runs in the spring and summer, the lions will follow any creek, canal, or riverbed, which travels through agricultural land for a water, cover, and an alternative food source.

The Ag land that lies on either side of the waterways provides a secondary type of hunting habitat from that of the rivers corridors. The "eat, hide," ritual of wildlife is still observed but, in a different setting than the forest or foothills. The type of crops grown dictates the percentage of cover.

Cover is linked to the kind of prey species' available, quantity of that species', and time of day or night when most prevalent. The valley has a diverse population of large to small rodents for mountain lions on the valley floor, when deer are not available. Though the river bottom is not included into my research, there is a deer herd that survives successfully there throughout parts of the year. Therefore lions will survive as well. In light of Ben's success at survival within the river region, and looking at the distance covered by waterways, it is my thought that there are more lion's than just Ben spending time at the Kings River.

Therefore, defining the, "fringes of society," is going to be a bit sketchy. When looking at the agriculture on the outskirts of towns, and water ways, I think that the question of, "How close do they really get to town," is going to depend on the location of the town, possibly the topography, and the town's population dynamics.

If I were to suggest a short management plan, it would be as follows. If a home owner can lock up their small stock like sheep and goats at night, it would save them the trouble of losing their animals and they'd get more sleep. Keeping a canine suited for small stock protection in with the animals would also be beneficial. For ranchers that have cattle, I would suggest purchasing a small herd of goats with two Billie's. They procreate pretty fast; the herd grows and keeps the lions from eating their calves…unless of course there's a habitual offender that moves into the area. From what I've seen with the number of lions passing through, the goats keep them happy for the majority of their range change. The number of mountain lions in a given area suggests a lesser or greater carrying capacity to support the number of mountain lions that find a niche in a given location. Mules work well for keeping lions at bay, they are temper mental! Putting dogs in kennels is expensive, but in the long run they are safer than on a cable. All enclosures need to have a roof on them, and be fully enclosed. Lions are great jumpers as I saw with Hookah. I think that the use of pesticides for burrowing rodents should be given consideration.

Attention should be focused on setting an amount that can be purchased by licensed persons such as a Qualified Applicator Licensed or a Qualified Applicator Certificate, and the locations where pesticides will be used. A concern for wildlife prey species' in conjunction with mountain lions travels is especially important in areas with low natural vegetation for which human travel is most prevalent in relation to public trails, and agricultural locations using pesticides. This thought should be seen as an option for public safety of people traveling alone on trails where less prey or alternative prey for the mountain lion can be found. Less food makes everyone a bit grouchier, and mountain lions more aggressive. Agricultural accountability in regards to pesticides not used in excess but properly, should keep burrowing rodents to a minimum yet not at a zero population.

This water holding pond is on Sneaker's route. The trees and grasses have grown up around it forming a small oasis that all wildlife uses for water, foraging, prey and cover. The weeping willow trees shade the banks in the summer.

People have invaded the lion's natural habitat and pushed the mountain lion down onto the valley floor. Now people don't know what to do other than, "Shoot the Mountain Lion," for trying to survive.

While conducting my research on the California Mountain Lion, I have learned many things. The most surprising is that the lions know how close we are in proximity to them. If they were so inclined, they could be more dangerous than I have found them to be. They do live a solitary life and though they live on our fringes, they generally leave humans alone, even when they accidently find one of us. It is of utmost importance to respect this species' space as they most often respect ours. When searching for a lion's story through track and sign, I have learned first-hand that I am responsible to "Never," forget safety or leave anything to chance. By observing from a distance and using common sense, hopefully I can continue to learn from these magnificent animals, and try to help preserve their future.

Chapter XXIX
Agricultural and Weather Effects on Mountain Lions

Despite all of the agricultural practices in the area, and the amount of people that live in some rural communities, our resident lions have done quite well maintaining their distance. This is considering rare events in the past with Barbara Alice, and her calf theft. It is up to people to make a difference. When I go into the field to research mountain lions, I am going into their habitat. I do not try to find a lion to see it up close, although sometimes it happens. I am interested in finding where they frequently go to explain how, and what they need to live in a given area. Finding where they den, rest, drink, and what food sources they are surviving on is important to see if, and for how long their seasonal habitat can support them, and how many of them it can support. Small game of squirrels, rabbits, and water fowl are indicative of remaining prey availability, even when the time comes for their seasonal range change.

When mountain lions are killed, their habitats become empty creating a hole for new mountain lions to move into. The increase in habitat allows new lions to spend more time away from their original habitat on seasonal range change. Each lioness that has established denning in this agricultural area will return when their current cubs disperse. Upon the subadults' dispersal, they may overlap the habitat that they traveled with their mother. This increase in lion population pushes their mother back to her original habitat. She will then raise another litter of cubs, and teach them to travel the same ground. Each cub learns the neighborhood, and places to either frequent, or avoid. Human presence is sparse in some of the study areas but, present. The scent, sound or sight of field workers, neighborhood residents, domestic animals, vehicles, and farm machinery, imprints each young lion in the area. They learn which places they can travel more often during the day, without being seen.

The Kings River and the Friant Kern Canal are two sources of irrigation water for the Central Valley. They are connected to the lower foothill region by a labyrinth of irrigation ditches, and sloughs. The majority of a lion's travel in this study is through agricultural land near riparian areas that are connected by these waterways, periodic holding ponds, and springs. The dispersal of mountain lions is not limited to the eastern side of the Kings River watershed. One young mountain lion from this study, "Ben," found his way across the river, and has made agricultural land on the west side of the Kings River a part of his home range.

The San Joaquin Valley is noted for radiation fog. The term does not mean the fog is radioactive. It means that a set of weather variables produces the fog. My understanding of this fog is that warm days and cool nights produce humidity which produces morning dew. The ground retains the moisture because the days remain at a warm, even temperature. The days are not in the triple digits any longer, so moisture stays within the soil and there is a percentage of water that is not returned to the water cycle through evaporation. With early evenings being cool the fog develops overnight, and reaches its full potential in the early morning hours. It can be clear and cool outside when we go to bed but, when we get up in the morning there's usually a thin blanket of mist or patchy fog. The moisture collects and the following mornings are often foggier until the killer fog moves in.

There are some years that we do not have radiation fog. People in the San Joaquin Valley call it, "Killer fog," because of the multi-car pile ups on the freeways, and the crashes at busy rural intersections. Fall, and winter rains increase the ground moisture leading to thicker fog the following day. We all have to plan to leave earlier to get to work and school. Sometimes the visibility is so low that I have to put the day off until the fog clears. After the rain comes in the winter, the possibility of fog low to the ground or hanging overhead is a given. The lionesses that stay in this area during the late fall to early spring do not mind the fog, they use it to their advantage. Barbara Alice is a felid that spends her early mornings hunting in the mist.

The fog hanging at ground level keeps the air warm around the base of the trees. The fog that hangs at mid-center or just above the tree line traps the cold air at the base of the trees. If the air at ground level is cold enough, it can do damage to tree fruit. Most people that haven't dealt with the fog would think that fog blankets an area and increases the air temperature. On the contrary, cold air settles and warm air rises. The wind machines are programmed to turn on automatically by the thermostat located at the base of the tower, when the temperature at ground level reaches a low of

162

between 33° and 31°. The thermostat is set according to the ranch owners' discretion.

When there is fog, and the wind machines turn on, there is less threat of frost because the blades pull down the warm air when in rotation. The fan blades are about 23 feet long and the air that's circulated comes from about twice the length of the fans blades. There is a temperature difference of about 4 degrees warmer which keeps the tree fruit from freeze damage.

Propane powered wind machines keep citrus crops from freeze damage in the winter.

Though the two oldest lions in the study spend time here during the foggy season, they are not fazed by the sound produced by wind machines. The machines keep the air temperature warmer, which enables each lion to hunt rodents at lower elevations. The majority of human activity on citrus Ag land during the foggy season is from propane trucks, and ranch foremen.

The refueling of propane tanks that power most wind machines takes place even during the night, to ensure that the impending frost won't harm the crops. The early morning hours between midnight and five a.m. is the quiet time when the dew and thin fog develops. When the fog descends to blanket the ground, the machines don't turn on. The lions can hunt rabbits or other small nocturnal game in silence. If the fog doesn't reach the ground, the machines turn on, and the lions hunt the sloughs, vineyards, and vacant farmlands that are in crop rotation. The dullness of sound fostered by the fog intern creates a hunting advantage for the lions. The times of day and night in which they hunt are unchanged.

The agricultural practice of spraying pesticides must be done in order to keep insect infestations at a minimum. The use of low dose solid poisons for rodent control is also practiced. These regimens help to produce healthy fruit, and vegetables for human consumption. California is the leading state for produce in the United States. The produce is grown in mass quantity to be shipped across the nation to supply grocery stores chains. Our produce is also shipped to other countries. The United States Department of Agriculture monitors the need for insecticides along with the potency required to keep infestations at bay. With crop studies, more often fewer insecticides and rodenticides are prescribed. With the help from mountain lions, less rodent poisons in the form of grain or barley are needed.

In regards to the travel patterns of the two lionesses during the times that crops are sprayed, their travel through these areas along dirt access roads comes to a stop depending upon what types of pesticides are being used. If citrus and grapes are being sprayed with sulfur, they stop taking the route for about two to three days. If another more potent spray is used, it can interrupt their travel patterns for a week or more, until the spray on the ground dissipates in potency, or water hits the remainder of the chemical on the ground. Although most of the spraying happens at night, some spraying occurs around 6:00 a.m. when the weather is cool. Spraying at night is not only to reduce the risk of passing cars being accidentally exposed to chemicals but, because some sprays must be used in a certain temperature with no direct sunlight. The farmers have employees that spot for the spray rigs, day or night, for a further precaution which proves to be helpful. The noise from the spray rigs doesn't scare the lions away, it only alerts them as to which area they may need to avoid. The scent of the chemical being used is distributed throughout the air on light breezes. This strong scent of poison is a deterrent for them. Some sprays have a re-entry time of 12 hours and other's longer. Ironically, humans sometimes ignore this fact but, mountain lions seem to observe it. This is in part due to the residual scent.

9/28/2011- While walking in the pasture across the road, I saw some clumps of colored oats. My first thought was that it was a grain rodenticide in strychnine form. That type of grain must be put under ground. The color is an iridescent greenish-blue. This poison is now harder to get a permit for. I know because I called the Department of Agriculture to research which poisons the farmer retained permits for. The oats that I saw were an iridescent blue, a different color all together. It was not supposed to be clumped up in piles, but scattered. The name of the poison is Diphacinone.

The blue pesticide, Diphacinone, sitting outside a squirrel's burrow was improperly distributed. The County Department of Agriculture acted quickly to help remedy the situation.

A county inspector came out to investigate the issue. I told him that I was worried about the resident wildlife specifically a mountain lioness, dying from eating a bird that had eaten the poison oats. He told me the reason for the blue and red colors used with the oats. The birds will eat the red oats but not the blue oats and that red oats are half the strength of the blue oats. The blue oats have to be scattered on the ground. The red oats are put in open plastic troughs or pipe feeders. When small rodents digest this grain, they go underground to die. The Ag inspector told me that any oat or barley pesticide with a gray or black color is extremely poisonous. This form of poison will deactivate when morning mist, rain or when sprinklers hit it. The phosphate turns into a gas, and then dissipates. Permits are required for individuals using this type of pesticide, and they are kept on file.

My other question was; what if she eats a squirrel that has eaten the poison but hasn't died from it yet? The inspector's answer was; she would have to eat three of them that were packed full of the grain to feel sick at her stomach. The poison was not strychnine. I was pretty happy about that, even though there shouldn't have been any poison set out in that fashion. The proper storage of pesticides should be practiced as well as proper disposal of animal carcasses poisoned from pesticides. The County Department of Agriculture monitors the use of these pesticides closely.

Diphacinone is a one kill poison. However there are problems with larger species' ingesting this poison then poisoning their small offspring through milk transfer. I learned this through Dr. Barrett. There is growing concern about the use of poisons. More research is being done.

The cows would go out in the mornings, and graze this property. It concerned me. I asked what affect the poison would have on cows. The answer came back as; a cow would have to eat about a hundred pounds of the oats to get a nose bleed. This was of importance to me because I have two cows that welcome the chance for grain, oats, or barley, no matter what color it was. If the cows had fingers they'd be dangerous.

Heavy machinery for tree chipping is necessary due to restrictions on burning because of air quality in the valley.

Tracy and I decided to take down the fences. The farmer was going to plant the pasture land in the near future anyway. It is possible for the poison to make a lion sick at its stomach. I found the tracks of a lioness, and a cub on the pond bank. I also found adult scats along with scats from a cub. The one thing I didn't plan to find was a dried, regurgitated pile of the blue oats. It is possible that a coyote ate the poison, then threw it up but, highly unlikely.

They seem to know when substances are poisonous and avoid them. When a coyote finds a food they don't like or find unsuitable, they will bury it.

When discussing the lioness, with the inspector, he had mentioned to me that he'd just spoken to a nearby party that had heard mountain lion screams about two or three weeks prior. Well, that would be about the same time that I found a large amount of wild felid spray on a small tree at the back of the lake on August 12th, 2011. I was correct with the determination that Sneaker's tracks were really his, and that there was a possible female within study zone 2, looking for a mate.

While monitoring the subadult, Ben, I saw that a heavy equipment company had been chipping trees for disposal. They moved their equipment in the night before, and began chipping early the next morning. They finished about 11:00 a.m. Ben's hunting excursion was over by the time they started work. It's taken him ten days to make a full circle. Agricultural machinery doesn't bother Ben. As a natural born resident cat, he was imprinted by farming operations, and farm laborers.

I have noticed that when watching three mountain lions from different locations, they share one common habit. They prefer to walk on smooth surfaces like trails and access roads. For example; I've watched Hookah travel through the pecan grove in agricultural land on a smooth manicured surface. While in the Sierra National Forest, a mountain lion used smooth logging roads. Shasta, a mountain lion who is confined at Critter Creek also uses his paths in either direction for travel. All of the lions prefer a well-used smooth trail. This tells me that no matter if the trail is public or private; the mountain lions will use them.

When rodent prey is limited or hunted out, the public should only use trails in secluded areas if they are in a group, and especially if mountain lions are prone to visiting that habitat. I always look for small wildlife when checking an area for general health, and to assess whether or not it could be considered possible habitat where frequency of travel may occur.

It is not enough to gauge a given amount of acreage by fuel height or tree spacing to evaluate the type of wildlife that should be present. Quantitative proof of wildlife needs to be identified either by using random selection in a given area for estimation or an actual population count. Tracking wildlife to find out which species' is in a given area is helpful. The successful carrying capacity of various habitat types for wildlife depends on several factors. Giving consideration to the height and density of ground vegetation on grass lands would indicate an average wildlife height or size. Having prior knowledge of which species' of omnivores, carnivores, and herbivores are within a distinct climate or habitat type allows for a range of wildlife to look for.

Moderate, average, or sparse tree density, with regard to spacing and canopy cover, with or without shrubs, and height of grasses, are indicative to what possible wildlife may be in that habitat. After researching the area to define which species' are present, population densities can be assessed. Riparian areas or any location with dense vegetation accompanied by fallen debris, and remnants of logs, within proximity to an available potable water source, and proof of wildlife are indications of a healthy habitat that can support a diverse ecosystem. Densely overgrown habitat near water means the possibility of larger prey species' being available in different seasons. In locations where the grass is short with few trees, means that small rodent prey such as squirrels, should be available in numbers.

Mountain lions, bobcats, gray fox, and owls, along with a few coyotes are more than happy to keep the squirrel population in check for a natural grassland habitat. The issue comes when squirrels or other small rodents on grazing lands are adjacent to agricultural lands. Mountain lions traveling sloughs and waterways through cattle grazing lands depend on the squirrel population to make up for the lack of deer or pigs in the area. The occasional squirrel is not enough to feed a lioness with cubs.

The tradeoff comes when one meal is substituted by another. Cotton tail bunnies and quail make up the balance when there are less squirrels or other prey in an area. Quail stay in a family group as they wander pecking at the ground for food. They are quick but, not as quick as Hookah.

There are different reasons for mountain lions to suffer impact from their surroundings. People are the main reason. However, the one impact that I have found in the study is not from farm equipment operations but, from the use of a pesticide. The application of pesticides for the poisoning of squirrels results in one less prey species' for mountain lions in select hunting areas. The loss of small rodent species' such as squirrels, and or gophers by means of poison affects all of the animal's diets. The health of the wildlife food chain depends on a stable balance to sustain both carnivores, and omnivores.

During the past three years there have been points of time where an influx in the squirrel population has been noted on agricultural land near us. Each farmer who has a permit is allowed to purchase an amount of Diphacinone pesticide. The purchase of the pesticide is made by a licensed person who has passed a pesticide application exam. A farmer may possess this license or he may hire someone else who holds this license. Unfortunately, some farmers that have the license give an unqualified worker the poison to distribute, and possibly give them the wrong instructions on application. More kills more.

During the spring and summer of 2011, and 2012, Diphacinone was applied incorrectly to two different citrus farms in zone 1. After reporting the first incident, I checked more closely when placing IR cameras so see that it was applied as directed. In two instances it was not. When I noted the second incident, I chose to travel the property to scatter the already, lumped up poison all over the ground. I drive by this location every day, and was alarmed when I saw that it had been distributed in piles all along the fence line at the front door of every squirrel hole.

The consequence to the amount of pesticide applied is that when squirrels don't eat all of the poison because it's left in such large quantity, it leaves the opportunity open for a different burrowing rodent species' to eat it, such as the Burrowing Owl. The Burrowing Owl takes over the vacant squirrel holes to raise their families in. They are an endangered species' here in California. Their survival will be affected by the misuse of pesticides. Diphacinone comes in two potencies, and is used for squirrels that weigh up to about three to four pounds. Gophers will also eat it. With each seasonal range change of mountain lions alternating habitat zones, the prey base must be there or along the way to a different hunting ground. If a lion in captivity can survive and be healthy on five pounds of meat a day, a mountain lion in the wild can at minimum survive on at least two squirrels a day until nocturnal hunting hours when he can hunt more prey. A lioness supporting new cubs will need at least twice as much prey. As her cubs grow an increase of prey will be needed. With the reduction of natural prey, they will resort to eating more domesticated animals. Thought- I have not personally seen this, I think that if a human were caught alone in this case, it could be possible that a normally calm, elusive mountain lion, that may be unfamiliar with an area, may bring danger to the human.

While it is necessary from the agricultural stand point to reduce the rodent population for healthy farming and cost effectiveness, I think that the amount of pesticides purchased by any licensed person(s) should be revisited and reassessed in light of individuals who've discovered that the incorrect application of pesticides kills faster and in quantity, versus the proper application of scattering, and a slower more minimal death rate. The correct method of using Diphacinone allows for a prey change at a gradual rate that carnivores can adapt to, rather than an abrupt prey loss. The county is good about taking care of issues like this but, I am only one person. I am not able to check the use of every farm that has applied this pesticide. Quite frankly, it would be considered trespassing if I tried.

Hookah has shown me that in the habitat she chooses to roam, she is able to overcome this impact by selecting bunnies and quail to eat for smaller prey substitution when there aren't as many squirrels to choose from. But, Hookah is our only long term lioness. During a two year period, this basin may see a population increase due to cubs, and subadults moving through on their way to find their niches. Though the impact of fewer squirrels is apparent, it is not dangerous just for one lioness, her cubs, Barbara Alice, and Henry, because of the healthy ecosystem in the basin (This statement may be refutable in the near future due to loss of small prey base). Barbara Alice only visits on a bi-annual basis, Hookah's cubs disperse, and Henry will be off to find a larger home range in the near future.

The next carnivore in our area that outnumbers the mountain lion is the bobcat. They face the same issue as the mountain lion with loss of prey species'. The danger aspect will come when there are more mountain lions that move down to the agricultural lands because of over population by humans at the foothill level. There are only so many niches for transient cougars to claim. These animals will not get the best areas which mean these sections will be short on burrowing rodents, and larger prey from the cougars that have already passed through. A substitution food source will be found. There will be no one to fault but the people who took proper dispersal of poisons for granted. In the long run the burden is suffered by the mountain lion.

Chapter XXX
Addendum

I feel satisfied with the answers I've found to the study criteria I set forth. I will always have questions but, now it's a whole lot easier to find the answers. I know there will be answers I do not like, and answers that take a long time to figure out. With Hookah dead, I will have to rely on Celia, Yinka, and their brothers to help me with these questions. They will lead me to relevant information that could help them in the future.

Note Regarding 1/27/12: On June 4th 2012, I found the tracks of one large lioness, and one cub in zone 1A. I've found nothing else to identify the animal. It was a lioness because the size of the tracks far exceeds any bobcat.

Henry Update- I wonder if having no mother from a young age pushed Henry to sleeping in trees. It would be the most sensible thing to do. Most of the time when I saw Hookah with her cubs, they were under a tree on top of a rock, but then all of her cubs had their mother to protect them.

I found an answer to how a male and female find each other if the male seeks a female to breed with, when she's still got cubs. When Ben came through on his first seasonal range change looking for a mate, the only female of breeding age was his mother, and she still had two cubs. Since then, Clause and Celia have turned 18 months old, and gone their own way.

Note: I found Celia's tracks visiting behind my parents' home.

Hookah Update- Hookah ventured through here on her way to the river. She followed Ben's calling cards (scat and scent markings) to find him. She may have found him by now. I'll do some checking when the sun comes up, to see if the people on his route have heard anything. Hookah will either be here, or on her way back in a short amount of time. Then she will meander around until the time grows closer for her to den at one of her favorite spots.

Hookah Update- In the beginning, I wondered if Hookah would den here again. Then she came back thinking that she would but, humans had destroyed her pond ecosystem. She had her cubs close by, and then moved to raised them in what used to be Suka's habitat. The cubs dispersed at 18 months old. Hookah set out on a quest looking for a male. She has returned to her home range. I have noticed that across the road is a small section of flattened down dry grasses near the fence line closest to the road across from my yard.

Hookah Update- The dogs woke me up at 2:00 a.m. creening. I looked out the window but, couldn't see anything. There were no flashlights that worked, so I grabbed my camera. The flash on the camera lights up the creek bottom and shows exactly what's out there in pictures. Yes, I got a picture of two large, glowing eyes looking back at me from the tall dry grass in the bottom. She came home.

Hookah Update- There has been no creening from dogs, the night and early mornings have been silent. It's 4:35 a.m. The one coyote that bravely howled two nights ago has vanished. Probably out of fear.

Perhaps she ate him. I'll check the locations that she roams later in the day to see what she's doing. When I have questions as to when and where a lion will cross one of these access roads in the study, I rake a section of soil, and check it twice a day for one to two weeks. I journal all of the information, take pictures of some evidence, and note the directions they are coming from and heading to, which lion it is, how many cubs are with a lioness, distance between each lionesses travel as opposed to the other's last known location, and so on.

I have been thinking about something. Hookah's mother had she, and a litter mate on the mountain. Her mother visits this location but, it isn't the center of her home range. Hookah has remained, making this the center of her home range. She had two litters of cubs here. The first litter resulted in only one living cub, Ben. She took him to the edge of her range to finish his upbringing. When he was ready to go, he revisited her haunts, then left to the direction that she does not frequent. Hookah then had another litter, Clause, Celia, and Jeremy. When they were old enough to travel she moved them to another foothill to finish raising them where new habitat had opened up. The lioness and her cubs have been seen by others in a nearby area, which is similar in size, shape, and agriculture, to this one. The cubs have become subadults, and were still together as of a week ago. Clause will probably venture off soon to find his own space. However, there's still one subadult left...Celia. I think Suka's home range will become Celia's home range. Did Hookah plan this? Maybe she did, and maybe Hookah's mother planned her area for departure as well. In time, she may hunt the same ground with her mother in opposite directions, along the same patterns that Suka hunted with Hookah. After Celia has her first cubs, and they are mobile enough to move, she will choose an area on the fringes of her center of home range to take them to until they are ready to go.

Subsequently this place may also dictate the direction for their dispersal. Since poor Sneaker is dead, his home range may still be in part, open for Clause. If this is how it works with mountain lions, perhaps it is why their territories overlap (i.e. mothers and daughters, aunts and sisters). I saw this with Ben, and the way he revisited areas where he'd spent time with his mother, then the direction in which he traveled when he left the area.

They have all revisited their mothers' haunts and now have moved on in a direction that she does not frequent. From a human perspective, maybe it also ensures that the kids won't feel compelled to stay at home, and try to take over the parents' home range. Another way to think about it is: When I have had each of my children, they stayed in my room with me until I felt comfortable moving them into their own rooms without me having to worry so much. They grow up using their bedrooms as their center of home range. I give my kids their own rooms so that they don't try to take over our master suite. There is only so much room before it becomes over crowded. They hunt in my kitchen, and then when they're old enough they venture outside under my supervision. The same happens with Hookah's cubs under her supervision. When my kids have grown even older they are allowed to go outside alone, as with Hookah's subadults. Eventually my kids become more mobile and visit the places I have gone and travel the same roads. The same happens with Hookah's offspring, they visit her known haunts. Then eventually my kids move out but, then come back for a visit on a scheduled basis.

Hookah's offspring do the same. All of my kids disperse through the doors in the house. Each door has a specific location. Apparently, Hookah has door ways for her offspring's dispersal as well. Like people, mountain lions move. This may be because of economics. People move for jobs, lions move for food. The doorways in each domain may change in the direction they face. This leads to a direction in dispersal.

Nevertheless, Hookah is home. Our silent neighbor has been working nights to decrease my domestic cat population. My friend is aghast at the fact that I don't keep them indoors. Well, it's quite simple. I do not care for the smell of a litter box or kitty accidents, nor do I want cat hair in my house. Though I have a bunch of outside cats, they stay outside because I'm allergic to the musty smell they create. It is also my opinion that cats need their own space, and should have their outdoor life. They are free to roam and make their own decisions on where, and how to hide. We are just lucky that we have the room for as many cats as I have. The vast majority of them make it through the lioness's presence just fine. Most of them are porch dwellers because they know I hear what goes on outside at night, and depend on the porch light flipping on. Hookah doesn't like the porch light. She stays in the shadows.

Hookah is getting smarter. 10:04 p.m. - I had been talking with Bryan in the kitchen just before going to bed. I was cleaning up from dinner, and opened the door to give the left overs to the cats. I walked out the door, called the cats, they all flew up onto the porch, and snapped up their goodies. As I was feeding them I heard the distress call of a kitten, like it was stuck in the tree. The location it was coming from was at the end of the house, in the shadows. I have no kittens. She did a pretty good job this time. Though she's getting better, she's still grasping. She doesn't make people sounds like some mountain lions do. I called Bryan to stick his head out the door to listen. I said, "Oh no Bryan, it sounds like a kitten is stuck in the tree." He said, "Why is it stuck in the tree, should I go get it?" I said, "Bryan, we have no kittens, and the one you'll find weighs about a hundred and thirty pounds!" He just looked at me and said he couldn't believe how much she sounded like a little kitten. I told him, "I know but that's how she catches her dinner." She didn't bank on me opening the door so late at night. The cats just grabbed their food, and ran for cover.

This is the last Hookah update. There is so much more to tell, but I am becoming blind to my own journaling and am afraid I may not catch errors I've made until this book has been printed. I cannot give the day or date but, it is now 12:50 a.m., and 3 days after Hookah's last visit. At 12:01 a.m.: I was sleeping great for a change then out of nowhere in the night came this rotten coyotes call just off to the side of the driveway about three rows of trees out. I flew out of bed! "That darn thing is still alive?" In two seconds the dogs went ballistic with their barking. The coyote never said a word. It was that coyote! It almost immediately got silent. Then from behind the kennel in the horses pasture came a familiar voice. It happened so fast that I knew it was Hookah. She voiced the whine of a puppy. Just two whines, then silence. The dogs turned their direction so fast, they creened one time, then more silence. Hookah had closed in on that coyote. I knew it was her making the whining sound, and that she had not mimicked the sound of the coyote because of the distance between the two calls. She was trying to pull the coyote back down into the pasture. She was hiding in the draw made by erosion on the hillside, just behind the last dog's kennel. The cat was given away by the dog's creen. She was so close to getting the coyote back down the hillside, too. I wish they'd not said anything to tip the coyote off. It had no idea Hookah was stalking it or how close she was. The escapade was over at 12:19 a.m. when one of the cats came flying up into the yard. This probably eliminated one of her nine lives.

The porch dwelling cats usually hide when Hookah hunts. Not this morning, they were all on the porch watching to see what would happen. Interesting fact, Hookah came home on the 12th; it is twenty five days into her gestational period. She has been around here in this zone since she came home, give or take eight days to travel some other hunting places. The last note on association: She associates different sounds with the corresponding sounds she hears from each location. My son William had a Catahoula puppy in the back yard for two weeks. The puppy would whine during the day, and sometimes in the evening. Hookah remembered this. She used it to draw the coyote in thinking he too, would associate the sound with the location. It almost worked.

The next evening Hookah had been resting in a close location. As the evening turned into night fall, the sound of spray rigs filled the air. Four of them came down the road to work on a forty acre parcel. They began at 10:00 p.m., a part of her hunting hours; she had to go somewhere else. It seemed like a chain reaction with the spray rigs and farm work. The area had been so busy for a week. Another week went by and the nights regained their silence. No Coyotes, no spray rigs, no crickets. The crickets even know when she's here. It took about two weeks after the farm work was over for her to come back at night. She's a quiet animal, doesn't bother anyone. She keeps the coyotes moving, I'm glad she's home.

The most important thing to do for my family is to remind them of our resident cats' occasional presence, and to never let my little girl go outside without me checking the area first. My little one isn't allowed outside without us. It is sometimes difficult to help Hanny understand that there are really big cats that she shouldn't be near, especially when she sees little fuzzy ones running around in the yard. I find that a documentary with large cats for her to see me watching works out well. She may run around while it's on TV but, she asks questions, sees how big they are, and how they eat. The rest will come.

I have made my youngest son very aware of what the mountain lions do. He has been tracking and researching with me for at least four years now. He's seen Ben, the youngest lion, while in the truck with my father. Bryan knows to feed the animals early, and is in the house before dusk. We make sure to store our garbage with the can lids closed. All of the pet's food is stored in closed containers with lids. My family and I spend a lot of time outdoors. Talking with my kids about the resident wildlife is something big in this house, and children need supervision. If not from wayward animals, then for the occasional rattle snake. From my first encounter with a mountain lion looking into my French doors, to see what I was doing, I've learned that the smell of food cooking or on the grill, having livestock, and small pets, along with the sounds of a small noisy child, "they will come."

Since Hookah likes to hide in the shadows, I am having a couple of lights installed so that there will be no surprises. The extra light will also push her further from the house. I find that over time I've become so familiar with how my research species' hunts, eats, sleeps, rests, travels, and learns, that I am more aware of where they are, and what they're capable of doing. I do not white wash what they do, they are predators. They are beautiful, but that should not be a deterrent in the human eye for what they are. They are wild animals.

Today the California Mountain Lion has evolved into the magnificent apex predator species' within farming and ranch land, a carnivore, which so many people forget exists. Without them, a natural respect for what their role is in an ecosystem, and the noncompliance to responsibly coexist with them, we as humans will be causing more problems, and damage for the world we live in. Hookah will only be here for a short time. Then she will take her cubs and go miles away. At that point she and her cubs will be lucky if they survive.

If I were to share my thoughts with a student considering a wildlife career, I would probably say, "Keep an open mind about people when out doing research." Most of my people encounters are good, and some of them are pretty great. There are people that are actually interested and enthusiastic about any kind of wildlife, and we swap stories. I just don't swap anything current or having to do with my target species'.

In the beginning when I was learning to do the research, there were some visits with people that were slightly upsetting. There are always those who know exactly what I'm looking for, and ask me questions trying to lead me into giving up a cat. I can smell them a mile away now, and yet they think they're being secretive. This type of person used to frustrate me. Now, I find them amusing. I just can't wait to see how that special person will try to get information out of me. Sometimes when they are walking up to me to begin the conversation with a hello, I can't help but chuckle a little. I just wait. I'm polite while listening to their stories. Sometimes it's easier to find some common ground.

Knowing what fuels the persons inquisitive nature can help me understand why they want the information they do. I try to keep the conversation positive so that in the end we part company happy or on a decent basis, with no cats brought up. I never let the person, no matter whom it is, know where I'm looking. I take pictures of Red Winged Black Birds, and squirrels. Being vague or short only irritates people. I tell them the truth. I'm taking pictures of wildlife. They ask what kind. I tell them, "anything that moves," which is actually true because one animal follows the other. It's always good to know how many different species' there are in a given area, and how prevalent each one is. I am lucky that the whole time I've done research on the mountain lion, that I have only had a few visits with people that made me wonder about the human race.

Fortunately now, I can look back and laugh. Generally, I have found that most people have good intensions, and greater hearts. I go out to research, and photograph on a regular basis. I never know who I'm going to see so, I keep an open mind. People can be an interesting source of information, and I always get a laugh out of a funny story. But in the end, when the sun rises again the next morning, my first stop before making coffee is to look out the front window in rare hopes of seeing a mountain lion. It's always a gift to see one.

In retrospect, things have changed from the beginning. When I started looking for a mountain lions trail, I was so new at this. I've made many mistakes. I have also learned from those mistakes. I am not a hunter. I've never been taught to stalk my prey. I had to learn from the mountain lion, how to find a mountain lion. There has been no one to hold my hand through this learning process. I never used any dogs or carried a gun. The mountain lion taught me to use more caution than I'd ever used throughout my life before. I've become more aware of my surroundings, and know that I am never alone. I've found that I have become more like the species' that I track. Each lion and lioness through their track, sign or visual has taken me with them, so to speak, on their journeys whether it be in their shadows or alongside them. I feel lucky that they've allowed me, from afar, to be with them to see how they move from one location to another on their meandering paths. I have learned from other species' how to tell when the mountain lions that travel through, are near. Dogs, horses, cattle, domestic cats, squirrels, crickets, birds, and even the coyote, through their habits have a way of alerting me to the presence of a mountain lion. Sometimes even the slightest sound that breaks an ominous silence can be attributed to the presence of a mountain lion. When looking for the presence of this large wild felid, my senses are heightened. I seems like I not only feel the air that flows over my skin but, smell it with all that it encompasses. I now see with more than my eyes. There is another world, a secret world, one which is only seen by few.

End Note: While education on wildlife is important, my real learning came from the species' *Puma concolor*, itself. It is my sincere hope that this book is used to educate people with what I have learned about mountain lion behavior, how humans and mountain lions can, and do coexist within a ranching and farming community unbeknownst to most of the human residents.

Considering the impacts on mountain lions that people normally would not think of, and how the effects on them can be prevented is a start for their preservation. Patience on the part of the human

179

is necessary to evaluate the reasons why a big cat may pay a short visit to a resident's home or property is needed. The time honored profession of ranching leads me to hope there can be more ranching preparedness in the future, with construction of enclosed shelters for small stock, and the possibilities of either introducing alternative prey sources for larger stock owners or further reducing the amount of pesticides used in either ranching or farming operations to balance out the smaller prey species'.

The willingness to understand why wildlife behaves as it does concerns all people. Humans regard themselves as being at the top of the food chain when essentially; we humans share this planet with the mountain lion and other large carnivores. The mountain lion's adaptation to agricultural farm land and operations is to be applauded considering they have not become extinct with what humans have done to their natural habitats, pushing them into the Ag lands. Learning from them will increase human awareness about how to keep both lions, and people safe from each other. Though man cannot always see them, they watch him, and learn! They survive, but not without great sacrifice. It is the responsibility of all people to ensure the safety of the Mountain Lion. When one of us goes, the other may follow; it's just a question of time. With regards to the big picture, the parallels of our lives are so similar.

While I am absorbed in researching mountain lions in their own wild habitat, I visited a mountain lion at Critter Creek. I had the honor of taking his picture for the book cover. When I got to his enclosure he met us with a tiny me-ow. It amazes me that such a small innocent sound can come out of such a large cat. I am used to their hunting sounds. Until Hookah made her distressed kitten call, I'd never heard a wild mountain lion meow. Shasta's human sibling, in relation to his place in their bond, describes the sound he makes as a peep-peep. I can't begin to imagine how these peoples souls were uplifted when they touched him as a little cub. How special that must have been. As Shasta's story goes, he was too imprinted by humans when Critter Creek got him, to be released again. At that time it was not allowed to rehabilitate a mountain lion in the state of California but, now it is given the right circumstances. Shasta's attitude and demeanor was enlightening to me. I had thought that if a cub were raised in captivity, it would have let the predatory nature go. I found that as they grow and mature in captivity, the nature remains even though it may not be acted upon. There was something about the first meeting with a mountain lion that healed my heart when Suka looked into my eyes. Her gaze brought the wonder back into my life. Though my first visit was with Suka, Shasta touched me as well. They speak without words, and I can bank on what they mean. I appreciate this, and have the utmost respect for what they embody. As magical as they are," they are, what they are." A mountain lion's distance whether in the wild or captivity, must be respected.

Shasta taught me something that I had been wondering about for the past few years. I had always seen pictures of mountain lions with blue eyes, and some with yellow eyes. I had thought at one point that the color of a lions eyes depended upon genetics. I then thought that maybe it depended on the elevations they lived at. When I first saw Shasta his eyes were blue in the shade. I looked at him again in partial sun and his eyes were slightly hazel with a hint of yellow. In the photo on the cover, his eyes are definitely yellow. He is in prowl mode. Genetics and elevation have nothing to do with it. It's because their eyes are mirrored to reflect the light they are in. They change with the light and what the lion is doing also seems to reflect in the color of their eyes. This would also explain why, when I get pictures of a lion's eyes at night, in their own habitat; sometimes they have a reddish glow to them. The reflection shifts from either greenish yellow to a hint of red around the edges.

Recently I got a picture of Hookah hunting at night. The only thing that I could see that didn't blend into the dry grasses at night, was the color glare of her eyes. While Shasta was in the shade, his eyes looked bluish gray. Their eyes are part of their camouflage. In the next picture, Shasta's eyes are almost a deep blue. His best friend took his picture. There is a great deal of familiarity there. I used this picture, and changed the background to a different habitat. I wanted to use it for the cover but, thought it might send the wrong message. I don't want people to think they are this beautiful, and docile looking all of the time. They are wild. When I was close to Shasta's enclosure, he looked me in the eyes. His eyes were a hazel color. He was assessing me, smelling the air in my direction. He didn't know who I was yet; he sat there and watched me anyway. He had decided he wanted to meet me. It was a nice meeting for two strangers but, it didn't feel like we were strangers. When he began walking his paths, surveying his territory, his eyes turned yellow in the sun. Shasta answered a question I'd had for a long time. I tried to do some research for the answer but, never found anything until I met him. Suka and Hookah's eyes were always yellow because they were in the sun, when I saw them clearly. When their eyes change color, it helps them hide in the open.

I also got to look closely at Shasta's facial markings. They are different from the mountain lions in my study zones. All of their facial markings are different in respect to how their eye liner defines their features. Not all mountain lions eye brows are the same color or shape. These facial markings are the best way I have found to tell them apart aside from the sizes of their tracks. Sometimes a lioness has a cub with very similar facial markings, yet there are still physical differences.

When I visited Critter Creek, I saw volunteers that have dedicated vast amounts of their time to the health and care of injured or orphaned birds and animals. The extensive knowledge they have gained through education and personal experience from a species' that has chosen them, is an asset to humanity. I saw that every volunteer acts with diligence to bring education and safety to the facilities visitors during open house. They have each helped with a project or the specific goals of the wildlife station to rehabilitate injured animals that qualify for future release back into the species' appropriate habitats. These people act anonymously with no expectations of any public recognition. They proceed with nurturing hearts to care for seized exotics unable to be released, and injured or orphaned birds and animals that would otherwise have met with an unhappy ending. These men and woman each share in a vital part of wildlife conservation for the sake of wildlife, and for future generations to come. I admire their conviction to each species' they are faced with helping. When listening to their stories about a particular resident animal, I laughed with them concerning the animal's disposition, what they had taught them, and how they'd either bonded or learned to work together. Each animal has its own personality and offers their individual perspectives during interaction. This is a unique place where animals are not only cared for and heard but, understood.

Their overall relaxed behavior and their eyes tell how they feel about where they live. They say they are happy to have been given a second chance at life.

Photo of Shasta, "The Mountain Lion," at Critter Creek Wildlife Station. Shasta is a lucky boy. He's very loved and cared for by his people. It shows in his eyes.

Many people are busy within their day to day lives. It is easy to think that someone else will be there to take care of the wildlife. It is my hope that more people will support wildlife rescues, and big cat sanctuaries. Donations to Critter Creek Wildlife Station can be made by visiting this web page. Please visit Critter Creek Wildlife Station at:

www.crittercreek.org/

A side view photo of Shasta hiding in his sanctuary. His profile shows his facial markings.

Shasta saunters over to see who I am.

From the Authoress

Thank you for reading "The Shadows of a Mountain Lion". I am my roughest critic. While researching, writing this book, and doing our first fund raising auction to donate (all auction) proceeds to Critter Creek, their mountain lion "Shasta" passed away. Many people were very sad as he was a miracle to so many.

Since writing this book I have learned so many new things about the mountain lions. During the past four years, I have been compiling information on the mountain lion's nocturnal hunting habits in our area. Currently, I am working on a new book titled "Nocturnal".

Donations to Critter Creek Wildlife Station can be made by visiting this web page. Please visit Critter Creek Wildlife Station at:
www.crittercreek.org/

References

Fjelline, D.P. and Mansfield, T.M. (1989). Method to standardize
the procedure for measuring mountain lion tracks. In Smith, R.H.
(Ed.). *Proceeding of the third mountain lion workshop. 1988,
Dec.6-8. Prescott, AZ. Arizona Game and Fish Department.*

Lexis Nexis Gould Publications. (2007). *California fish and game
code handbook.* (2005-2006 ed., p. 313 SS4800). Charlottesville,
VA: Lexis Nexis Gould Publications.

Shaw, H.G., Beier, P, Culver, M., & Grigione, M. (2007). *The
cougar network presents puma field guide: In a guide covering the
biological considerations, general life history, identification,
assessment, and management of Puma concolor* (85 ed.). The
Cougar Network 2007. Retrieved from http://www.cougarnet.org

U.S. Department of Agriculture, Forest Service (1995). *American
Marten, Fisher, Lynx, and Wolverine: Survey Methods for their
Detection (PSW GTR-157).* Washington, DC: U.S.Government
Printing Office.

www.ingramcontent.com/pod-product-compliance
Lightning Source LLC
Chambersburg PA
CBHW051910170526
45168CB00001B/322